Kinetics Applied to Organic Reactions

Studies in Organic Chemistry

Executive Editor
Paul G. Gassman

Professor and Chairman
Department of Chemistry
University of Minnesota
Minneapolis, Minnesota

Volume 1 Organoboranes in Organic Synthesis
Gordon M. L. Cragg

Volume 2 Molecular Orbital Methods in Organic Chemistry HMO and PMO: An Introduction
William B. Smith

Volume 3 Highlights of Organic Chemistry
William J. le Noble

Volume 4 Physical Organic Chemistry: The Fundamental Concepts
Calvin D. Ritchie

Volume 5 Adamantane: The Chemistry of Diamond Molecules
Raymond C. Fort, Jr.

Volume 6 Principles of Organic Stereochemistry
Bernard Testa

Volume 7 The Alkaloids: The Fundamental Chemistry–A Biogenetic Approach
David R. Dalton

Volume 8 Concepts of Organic Synthesis: Carbocyclic Chemistry
Bradford P. Mundy

Volume 9 Kinetics Applied to Organic Reactions
Wiendelt Drenth and Harold Kwart

Other Volumes in Preparation

Kinetics Applied to Organic Reactions

WIENDELT DRENTH
Laboratory for Organic Chemistry
University at Utrecht
Utrecht, The Netherlands

HAROLD KWART
Department of Chemistry
University of Delaware
Newark, Delaware

MARCEL DEKKER, INC. New York and Basel

Library of Congress Cataloging in Publication Data

Drenth, Wiendelt.
Kinetics applied to organic reactions.

(Studies in organic chemistry ; v. 9)
Includes index.
1. Chemical reaction, Rate of. 2. Chemistry, Organic. I. Kwart, H., [Date] joint author. II. Title.
QD502.D73 547.1'394 79-27452
ISBN 0-8247-6889-2

MARCEL DEKKER, INC.
270 Madison Avenue, New York, New York 10016

Current printing (last digit):
10 9 8 7 6 5 4 3 2 1

PRINTED IN THE UNITED STATES OF AMERICA

Preface

The concept which sustained the effort to put this book together was born of an urgent necessity--one that is coming sharply into focus at many universities where the numbers of graduate students in organic chemistry have seriously declined in recent years. It is now widely recognized that many graduate courses of considerable interest to organic chemists cannot be taught in the traditional way, i.e., by a lecturer meeting with a class consisting of a minimum number of students. In the large majority of graduate schools of chemistry this minimum-number requirement--usually about six students, to be consistent with even poor teaching economics--cannot be met more frequently than one semester in two to four years. The result is that a good many students graduate without ever having been exposed to instruction in the ancillary subjects deemed to be of fundamental importance to the well-rounded researcher in organic chemistry (and biochemistry); in our opinion kinetics in mechanistic organic chemistry is one such subject that for this reason has often been neglected in the curriculum vitae of the modern organic chemist's career.

It is our contention that under the prevailing circumstances a tutorial system offers a viable alternative to customary classroom lecturing. Indeed, it may even be superior as a way of maintaining the continuity of the educational process while the organic student is deeply immersed in the preparation of the laboratory part of his dissertation. If a tutorial is built around a well-written book of carefully delineated scope, even a good lecturer in the classroom can be displaced without sacrifice of the basic course values. The crucial proviso, however, is that the book be more or less self-teachable; that is, the student can master the material with a minimum amount of help from the tutorial instructor. This is to be seen as an efficient and economical process from the point of view of both the instructor and the university's teaching resources. The instructor spends considerably less time in the course of an assigned tutorial, meeting occasionally with one or two students, than he would be required to devote to preparing for and meeting a class (say) three hours per week. The student, proceeding at his self-chosen pace, meets with the instructor only when he has some questions to be clarified by personal discussion and finally takes an oral exam pitched to test a predetermined level of accomplishment on the student's part.

There is another audience to which this book is also addressed, comprised mainly of practicing organic and biochemical researchers who have experienced a desire to review the subject matter or to update a background in mechanistic studies which was somewhat neglected in professional training of an earlier vintage. A volume designed to fill the requirements of a tutorial teaching relationship and relying largely on the student's initiative in the learning process should accommodate most readily the demands of this group.

It was with such objectives in mind that we faced the task of bringing the book to publication. The actual text has grown out of experiences of many courses in kinetics for graduate students in organic chemistry and biochemistry presented at our respective universities. These courses were intended to teach how to apply the kinetic tools in elucidating reaction mechanisms. Thus, the emphasis is on application because this is most consonant with the needs of the practicing organic and biochemical researcher, rather than on theory, although theory has by no means been neglected. Among a large number of authoritative books on kinetics almost none are aimed exclusively at these goals. Here, isotope effects are treated in relatively greater depth because it is felt that such kinetic considerations will play an increasingly important role in future studies of reaction mechanism. The chapter on chain reactions includes discussions of such matters as the rotating sector method for determining absolute rates, which is considered of essential importance to subjects such as polymer chemistry, photochemistry, and the mechanisms of life reactions. Strangely enough we have found that the more general textbooks on kinetics do not provide treatments of many subjects which are deemed to be of highest potential importance to the audience we wish to reach.

To a preponderant extent we have used the International System of Units based on our belief that the student of today will have urgent need to be familiar with this system tomorrow.

We have profited from the remarks and suggestions solicited from many colleagues and co-workers. We are particularly grateful to Dr. J. F. Arens, Dr. R. J. M. Nolte, Dr. W. F. Verhelst, and Mr. A. Boelee.

Wiendelt Drenth
Harold Kwart

Contents

Kinetics Applied to Organic Reactions

1 Introduction

Where do kinetics fit into the study of organic chemistry?

The study of chemistry may be separated into several categories:

1. Chemical conversions, i.e., dynamic phenomena
2. Structure analysis, i.e., characterization of a given molecular state (this division applies not only to chemistry as a whole but also to organic chemistry to which the scope of this book will be confined)

Structure analysis as currently practiced is based mainly on spectrometric techniques such as magnetic resonance, infrared absorption, x-ray diffraction, etc. In the most recent decades the empirical basis of chemical conversions, as used originally, has been replaced by an approach in which a full comprehension of the reactivity of the various classes of compounds is the designated objective. In other words, realizing a clear perception of the course of a chemical reaction and knowing how and why a chemical reaction will proceed, are of prime importance.

$$A + B \longrightarrow C + D$$

KINETICS VERSUS THERMODYNAMICS

In principle, every reaction can be said to proceed. It is only a matter of to what extent and how easily. Thermodynamics provide an answer as to the possible extent of occurrence of a reaction through consideration of such problems as the reaction equilibrium and the Gibbs function. If under a given set of conditions the Gibbs function of C + D is much smaller than that of A + B, at the stipulated temperature and pressure the equilibrium

$$A + B \rightleftarrows C + D$$

lies far to the right. In other words, if we bring together compounds A and B, they will react with each other until the position of equilibrium is reached. Under this set of conditions there will be practically a quantitative conversion of A and B into C and D. In the reverse case starting with C and D, the system will reach the same position of equilibrium, but the total conversion will be very small.

For example: If we add 1 liter of 2 mol/liter HCl to 1 liter of 2 mol/liter sodium acetate, a near quantitative conversion to 2 liters of 1 mol/liter un-ionized acetic acid will be formed along with sodium and chloride ions. The equilibrium

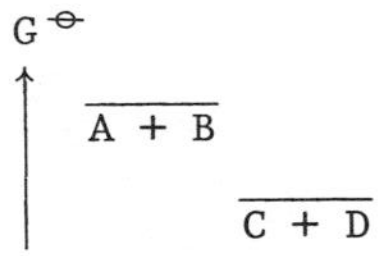

$G^{\ominus}$ is the Gibbs function, also called Gibbs free energy or simply free energy (F).

$$H^+ + CH_3COO^- \rightleftharpoons CH_3COOH$$

lies far to the right. The Gibbs function of undissociated acetic acid is much smaller than that of a mixture of protons and acetate ions.

Another example: If tert-butyl chloride is dissolved in, e.g., dimethyl sulfoxide, it can ionize to a carbocation and a chloride ion:

$$(CH_3)_3CCl \rightleftharpoons (CH_3)_3C^+ + Cl^- \qquad (1.1)$$

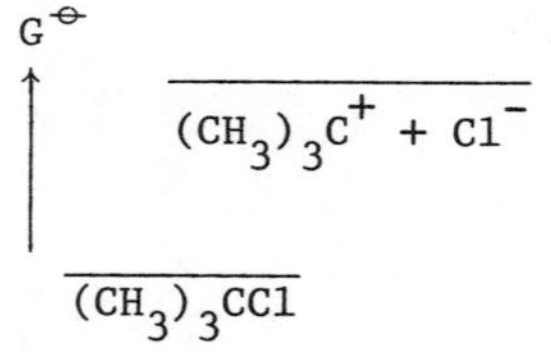

The carbocation is such a highly energetic particle and the Gibbs function of $(CH_3)_3C^+ + Cl^-$ so large that the position of equilibrium lies far to the left. The total amount of carbocations is very small and cannot be detected by ordinary spectrometric methods. The situation changes when water is added to the solution. The cation reacts very rapidly with water:

$$(CH_3)_3C^+ + H_2O \rightleftharpoons (CH_3)_3COH + H^+ \qquad (1.2)$$

The position of this equilibrium lies extremely far to the right. The tert-butyl cations are removed, equilibrium (1.1) shifts to the right, and the tert-butyl chloride is completely converted for all intents and purposes.

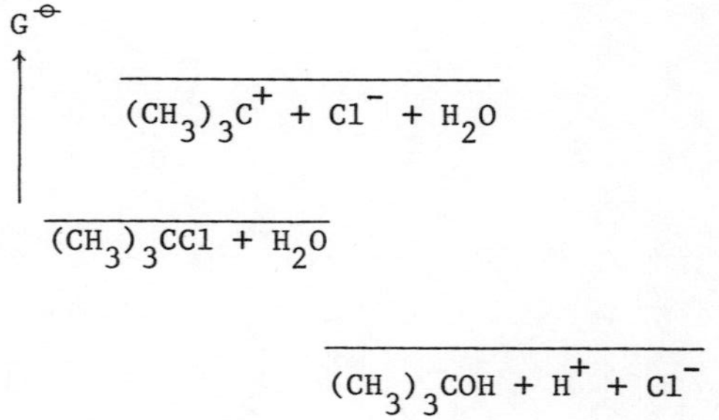

At the same time, this example illustrates that it is often possible to distinguish different steps in a reaction process, in this case (1) and (2):

(1) $(CH_3)_3CCl \rightleftharpoons (CH_3)_3C^+ + Cl^-$

(2) $(CH_3)_3C^+ + H_2O \rightleftarrows (CH_3)_3COH + H^+$

The separate steps are called elementary steps.

If a reaction can be said to proceed in the thermodynamic sense, this is not to say that it does so with a noticeable rate. In this book we shall be mainly concerned with rates of reactions.

IMPORTANCE OF QUANTITATIVE DATA

In preparative chemistry reaction rate is usually the most important factor. In undertaking the synthesis of a compound C from A and B, if under the specified conditions of pressure and temperature the reaction does not proceed to a significant extent, we may surmise that the reaction rate is too low under these circumstances. Another possibility can occur (say) when we try to prepare compound L from compound K but compound M is formed instead of compound L. This behavior is a matter of relative reaction rates; either the reaction rate K → M is much higher than the reaction rate K → L, or L is formed first and is converted rapidly into M. The formation of M is facilitated when, for example, it precipitates.

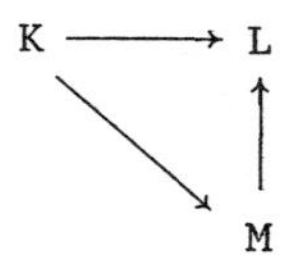

These are only qualitative reflections on reaction rates. Collecting quantitative data requires more effort but is often more instructive as to the reaction mechanism and therefore more rewarding. A good illustration is the classical study of nucleophilic substitution by Hughes, Ingold, et al. Their work represents a milestone in organic chemistry. It is concerned mainly with the determination of rates and the formulation of kinetic generalizations, correlated with reaction stereochemistry.

The decomposition of an alkyl bromide in 80 vol % alcohol under the influence of 0.01 mol/liter of NaOH at 55°C serves as a good illustration:

$$RBr + OH^- \longrightarrow ROH + Br^-$$

The rate equation is $v = kc_{RBr}$, where v is the rate, k the rate constant, and c_{RBr} the concentration of the alkyl bromide. The following values for the rate constant were observed:

R	Me	Et	i-Pr	t-Bu
$k \times 10^5\ s^{-1}$	21.4	1.7	0.29	1010

The sudden, great change in the reactivity series in going from isopropyl to tert-butyl suggests that a change in reaction mechanism has occurred. It is recognized now that these mechanisms are S_N2 and S_N1, respectively.

Another example is the halogenation of acetone catalyzed by a base:

$$CH_3-\overset{\overset{\displaystyle O}{\|}}{C}-CH_3 + X_2 \xrightarrow{OH^-} CH_3-\overset{\overset{\displaystyle O}{\|}}{C}-CH_2X + HX$$

The rate of halogenation is observed to be independent of the concentration of the halogen but is dependent on the concentration of the

C. K. Ingold, Structure and Mechanism in Organic Chemistry (Cornell Univ. Press, Ithaca, N.Y., 1953), p. 318.

The rate or velocity is the change in concentration during a certain time interval divided by that interval. In this example

$$v = \frac{dc_{ROH}}{dt} = \frac{dc_{Br^-}}{dt}$$

$$= -\frac{dc_{RBr}}{dt} = -\frac{dc_{OH^-}}{dt}$$

base. These observations illustrate that it is not necessary for each compound taking part in the reaction stoichiometry to appear in the rate equation. They are consistent with the following steps:

$$CH_3-\underset{\underset{\text{I}}{\|}}{\overset{O}{C}}-CH_3 + OH^- \rightleftharpoons \left(CH_3-\underset{\underset{\text{II}}{\|}}{\overset{O}{C}}-\bar{C}H_2 \longleftrightarrow CH_3-\underset{|}{\overset{O^-}{C}}=CH_2 \right) + H_2O$$

$$II + H_2O \rightleftharpoons CH_3-\underset{\underset{\text{III}}{|}}{\overset{OH}{C}}=CH_2 + OH^-$$

$$III + X_2 \rightleftharpoons \left(CH_3-\underset{\underset{+}{|}}{\overset{OH}{C}}-CH_2X \longleftrightarrow CH_3-\underset{\underset{\text{IV}}{\|}}{\overset{\overset{+}{OH}}{C}}-CH_2X \right) + X^-$$

$$IV + H_2O \rightleftharpoons CH_3-\overset{O}{\overset{\|}{C}}-CH_2X + H_3O^+$$

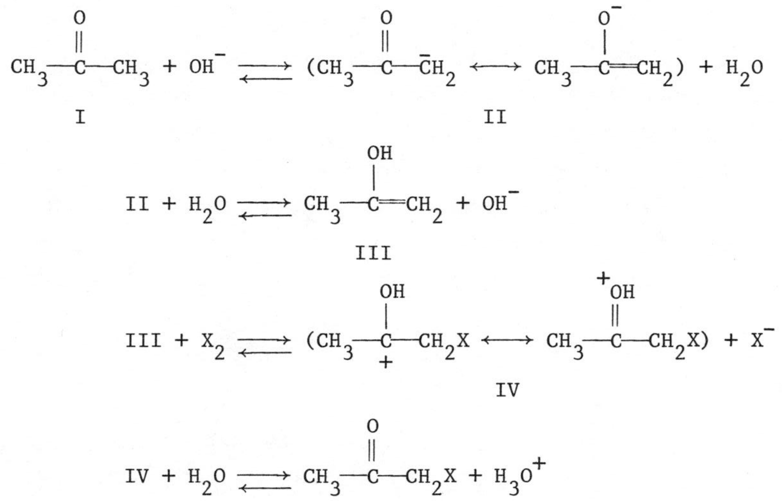

In principle each step is reversible. It will also be clear that some knowledge of the chemistry is required in addition to the kinetic results. The first step is clearly the removal of a proton by the base participating in the reaction. A C—H bond is broken, and experience tells us that the ionization of a C—H bond is ordinarily a slow reaction as compared, for instance, to the ionization of an O—H bond. The first step is therefore identified as the slow one. The other steps are much faster. The overall rate of reaction is regulated by the

R. P. Bell, The Proton in Chemistry (Cornell Univ. Press, Ithaca, N.Y., 1973), P. 132.

We define ionization conventionally as the conversion of a covalency to an electrovalency, i.e.,

$$C-H \rightleftharpoons C^-H^+$$

slowest step, in this case the first step. The rate of this step is given by the kinetic expression

$$v = kc_{acetone}c_{OH^-}$$

where v is the velocity of reaction, k a proportionality constant called the rate constant, and c_X the concentration of each of the reacting species involved in this rate-determining step. This is exactly the equation which is verified by the experimental data.

The intermediate, the enol form, cannot be isolated because it is consumed by the halogen much faster than its equilibration with acetone. In kinetic terms

$$k_2 c_{enol} c_{X_2} >> k_{-1} c_{enol}$$

$$\text{acetone} \underset{-1}{\overset{1}{\rightleftarrows}} \text{enol}$$

$$\text{enol} + X_2 \underset{-2}{\overset{2}{\rightleftarrows}} \text{halogenide}$$

Since the halogen takes part in the reaction after the slowest step, its concentration is without influence on the overall reaction rate. The halogenation rate thus becomes equal to the rate of creating the enol form from the keto form. With halogen concentrations much smaller than normal the rate of reaction of X_2 with enol, i.e., step 2, becomes slower than the reversal of enol formation in step 1, and equilibrium is thus established between the keto form and the enol form of the ketone. Under these conditions the concentration of halogen appears in the rate equation.

These examples are meant to illustrate how organic chemical research is broadened when reaction rates are measured. The study of reaction rates and kinetic equations normally falls within the scope of physical chemistry. The application of physical chemistry, however, is an expedient and not an objective in itself for the organic chemist. The aim of the organic chemist in applying these physical methods is the study of reaction mechanisms. More detailed treatments can be found in

A. A. Frost and R. G. Pearson, Kinetics and Mechanism, 2nd ed., (John Wiley, New York, N.Y., 1961).

E. A. Moelwyn-Hughes, Chemical Statistics and Kinetics of Solutions (Academic Press, New York, N.Y., 1971).

C. H. Bamford and C. F. H. Tipper, eds., Comprehensive Chemical Kinetics (several volumes) (Elsevier, Amsterdam)

Other books are

K. Schwetlick, Kinetische Methoden zur Untersuchung von Reaktionsmechanismen (Deutscher Verlag u. Wiss., Berlin, 1971)

H. E. Avery, Basic Reaction Kinetics and Mechanisms (MacMillan, London, 1974)

A. Weissberger, Techniques of Chemistry, Vol. VI (Investigations of rates and mechanisms of reactions, Part I, 3rd ed., E. S. Lewis, ed., Part II, 3rd ed., G. G. Hammes, ed. (John Wiley, New York, N.Y., 1974).

Chemical reactions take place in either a homogeneous or a heterogeneous environment. Homogeneous corresponds to reactions occurring in one phase. The scope of this text encompasses mainly reactions in a homogeneous atmosphere. Homogeneous reactions are usually more susceptible to interpretation than heterogeneous reactions.

The most significant difference between reactions in the liquid phase and the gas phase is the contact time subsequent to a collision of the particles. In the gas phase two colliding particles separate almost immediately. On the other hand, in the liquid phase the molecules are kept together by a cage of liquid molecules, i.e., the cage effect. For example, the residence time (the time that two particles stay together) in a water cage has been estimated to be about 10^{-11} to 10^{-12} s. In this interval the particles collide with each other some 10 or 100 times. If two particles are so reactive that nearly each collision is fruitful or at least 1 in 10 collisions produces reaction, then the reaction rate is limited only by the rate with which diffusion brings the particles together. This behavior is called diffusion limited or diffusion controlled. Examples can be identified among reactions in which a positive and a negative ion or two radicals combine, for example,

$$H_3O^+ + OH^- \longrightarrow 2H_2O \qquad \text{(see Chapter 6)}$$

$$2I \longrightarrow I_2$$

J. Halpern, J. Chem. Educ. 45:373 (1968).

2 Kinetic Equations

Selection among the many possibilities

Some review of a few simple kinetic expressions is in order. Let us suppose compound A is converted into one or more products in an irreversible reaction and, furthermore, the rate of conversion of compound A is proportional to its concentration c_A. The reaction rate is written as $-dc_A/dt = kc_A$. The proportionality constant k is called the rate constant of the reaction.

$$\frac{dc_A/dt}{c_A} = -k \quad \text{or} \quad \frac{d \ln c_A}{dt} = -k$$

Integration:

$$\ln c_A = -kt + \text{constant} \tag{2.1}$$

At $t = 0$, the concentration has its initial value c_A^0, so that the constant of integration is equal to $\ln c_A^0$. Therefore,

$$\ln \frac{c_A}{c_A^0} = -kt \qquad \text{or} \qquad c_A = c_A^0 \exp(-kt) \tag{2.2}$$

$\exp a \equiv e^a$

If compound A is converted into compound B and the concentration of compound B has its initial value of zero, then the concentration of compound B is given by

$A \longrightarrow B$

c^∞ is concentration at infinity.

$$c_B = c_B^\infty[1 - \exp(-kt)]; \qquad c_B^\infty = c_A^0 \tag{2.3}$$

$c_A^0 \longrightarrow \frac{1}{2} c_A^0$

These equations are also applicable to the decomposition of radioactive isotopes. The half-life, $t_{1/2}$, is the time interval in which the initial amount of the isotope has decayed to half its original value:

$$\frac{1}{2} c_A^0 = c_A^0 \exp(-kt_{1/2}) \quad \text{or} \quad kt_{1/2} = \ln 2$$

$$k = \frac{\ln 2}{t_{1/2}} = \frac{0.693}{t_{1/2}} \quad \text{and} \quad t_{1/2} = \frac{0.693}{k} \tag{2.4}$$

Half-lives of isotopes:

$T \equiv {}^3H = 12$ y

${}^{32}P = 14.3$ d

${}^{14}C = 5568$ y

For each segment of the curve a in Fig. 2.1, during a time interval $t_{1/2}$, the concentration is reduced to half of the initial value. In common logarithmic form the equation expressing this is

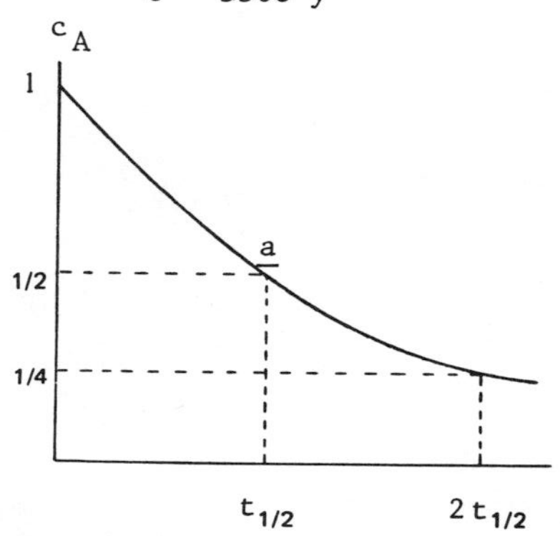

FIG. 2.1

$$\log c_A = -0.434kt + \log c_A^0 \qquad (2.5)$$

$$\log c_A = 0.434 \ln c_A$$

FIRST ORDER

Plotting $\log c_A$ versus the time gives a straight line (Fig. 2.2). The slope is -0.434k. Reactions which obey this equation are first-order reactions. The dimension of their rate constant is $time^{-1}$; k is expressed in the units s^{-1}, min^{-1}, or h^{-1}. The unit s^{-1} is preferred because the second is the unit of time in the SI system.

First-order kinetics are also applicable to the transition of an excited state to the ground state. For anthracene the rate constant for the transition from the first excited singlet to the ground state is $k = 6 \times 10^7\ s^{-1}$, and that of the first excited triplet to the ground state is $10^4\ s^{-1}$. The reciprocal of k is called the lifetime of the excited state, $1/k = \tau$. In the latter example the lifetimes are 16.7×10^{-9} s and 0.1×10^{-3} s, respectively. The huge difference is due to the fact that the ground state is a singlet and, in agreement with quantum mechanical considerations, the singlet-singlet transition is allowed, while the triplet-singlet transition is, to a first approximation, forbidden. Accordingly, the triplet state has a longer lifetime than the singlet state.

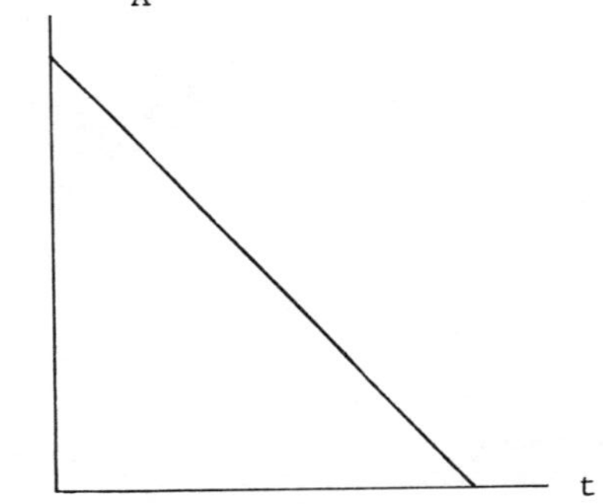

FIG. 2.2

N. J. Turro, Molecular Photochemistry (Benjamin, New York, 1967), pp. 49-50.

FIRST ORDER WITH A REVERSE REACTION

The problem is a little more complicated when the reaction is reversible. Assume that we start with A alone: c_A^0. After an appropriate time interval equilibrium will practically be reached. Designating the concentration of A as c_A^∞ and that of B as c_B^∞ at equilibrium (see Fig. 2.3),

$$A \underset{k_{-1}}{\overset{k_1}{\rightleftarrows}} B$$

$$k_1 c_A^\infty = k_{-1} c_B^\infty$$

The equilibrium constant is

$$K = \frac{k_1}{k_{-1}} = \frac{c_B^\infty}{c_A^\infty}$$

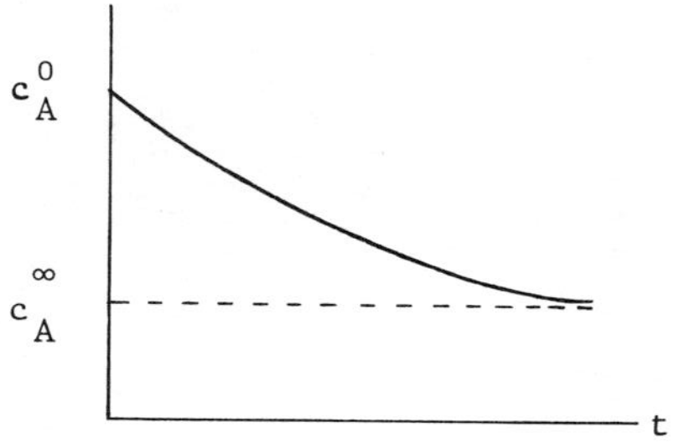

FIG. 2.3

K can be calculated from the final state composition. At any moment

$$c_B = c_A^0 - c_A$$

$$\frac{dc_A}{dt} = -k_1 c_A + k_{-1} c_B$$

$$\frac{dc_A}{dt} = -k_1 c_A + k_{-1} c_A^0 - k_{-1} c_A$$

$$= -(k_1 + k_{-1}) c_A + k_{-1} c_A^0$$

$$= -(k_1 + k_{-1}) \left(c_A - \frac{k_{-1}}{k_1 + k_{-1}} c_A^0 \right) \tag{2.6}$$

$$\frac{dc_A/dt}{c_A - [k_{-1}/(k_1 + k_{-1})] c_A^0} = -(k_1 + k_{-1})$$

or

$$\frac{d}{dt} \ln\left(c_A - \frac{k_{-1}}{k_1 + k_{-1}} c_A^0\right) = -(k_1 + k_{-1})$$

Integration:

$$\ln\left(c_A - \frac{k_{-1}}{k_1 + k_{-1}} c_A^0\right) = -(k_1 + k_{-1})t + \text{constant}$$

For $t = 0$ and $c_A = c_A^0$, the integration constant is equal to

$$\ln\left(c_A^0 - \frac{k_{-1}}{k_1 + k_{-1}} c_A^0\right) = \ln\left(\frac{k_1}{k_1 + k_{-1}} c_A^0\right)$$

Thus,

$$\ln \frac{(k_1 + k_{-1})c_A - k_{-1}c_A^0}{k_1 c_A^0} = -(k_1 + k_{-1})t \qquad (2.7)$$

or

$$K = \frac{k_1}{k_{-1}}$$

$$\ln \frac{(K + 1)c_A - c_A^0}{Kc_A^0} = -(k_1 + k_{-1})t \qquad (2.8)$$

Because $c_A^0 - c_A = c_B$, the equation can also be written as

$$\ln \frac{Kc_A - c_B}{Kc_A^0} = -(k_1 + k_{-1})t \qquad (2.9)$$

When the initial concentration of B is not zero but c_B^0, this equation becomes

$$\ln \frac{Kc_A - c_B}{Kc_A^0 - c_B^0} = -(k_1 + k_{-1})t \tag{2.10}$$

Plotting the natural logarithm versus time gives a straight line with a slope equal to $-(k_1 + k_{-1})$.

Equation (2.8) can also be written as

$$c_A = \frac{1}{K + 1} c_A^0\{1 + K \exp[-(k_1 + k_{-1})t]\}$$

Since

$$c_A^\infty = c_A^0 - c_B^\infty = c_A^0 - Kc_A^\infty$$

then

$$(1 + K)c_A^\infty = c_A^0$$

By means of the latter relationship, Eq. (2.6) can be transformed into

$$\begin{aligned}\frac{dc_A}{dt} &= -(k_1 + k_{-1})c_A + k_{-1}(1 + K)c_A^\infty \\ &= -(k_1 + k_{-1})c_A + (k_1 + k_{-1})c_A^\infty \\ &= -(k_1 + k_{-1})(c_A - c_A^\infty)\end{aligned}$$

The rate is proportional to the distance removed from equilibrium, and equilibrium is approached with a rate constant equal to $k_1 + k_{-1}$.

In cases where the reverse reaction is negligible, that is, where $k_{-1} << k_1$ and $K \simeq \infty$, the logarithmic equation (2.8) can be rewritten as

$$\ln \frac{(1 + 1/K)c_A - (1/K)c_A^0}{c_A^0} = -(k_1 + k_{-1})t$$

This reduces to $\ln(c_A/c_A^0) = -k_1t$, which is equivalent to Eq. (2.2).

SECOND ORDER

When the reaction is second order in one component, the rate is proportional to c_A^2:

$$\frac{-dc_A}{dt} = kc_A^2$$

$$\frac{-(dc_A/dt)}{c_A^2} = k \quad \text{or} \quad \frac{d}{dt}\frac{1}{c_A} = k$$

Integration gives $1/c_A = kt + \text{constant}$.

When the initial concentration at $t = 0$ is c_A^0, then the integration constant is $1/c_A^0$, so that

$$\frac{1}{c_A} - \frac{1}{c_A^0} = kt \tag{2.11}$$

With the aid of Eq. (2.11) it is possible to calculate the rate constant k from concentrations c_A measured at various times t. Another method is plotting $1/c_A$ or $1/c_A - 1/c_A^0$ versus the time t. A straight line is obtained, the slope of which is equal to k.

The dimension of this second-order rate constant is concentration^{-1} time^{-1}. The preferred units are liter mol^{-1} s^{-1}.

When, at $t_{1/2}$, the initial concentration c_A^0 is halved,

$$\frac{1}{(1/2)c_A^0} - \frac{1}{c_A^0} = kt_{1/2} \quad \text{or} \quad \frac{2}{c_A^0} - \frac{1}{c_A^0} = \frac{1}{c_A^0} = kt_{1/2}$$

then

$$t_{1/2} = \frac{1}{kc_A^0} \tag{2.12}$$

Here the half-life interval is not constant throughout the course of reaction, as is the case in a first-order process, but is, in fact, inversely proportional to the initial concentration (see Fig. 2.4).

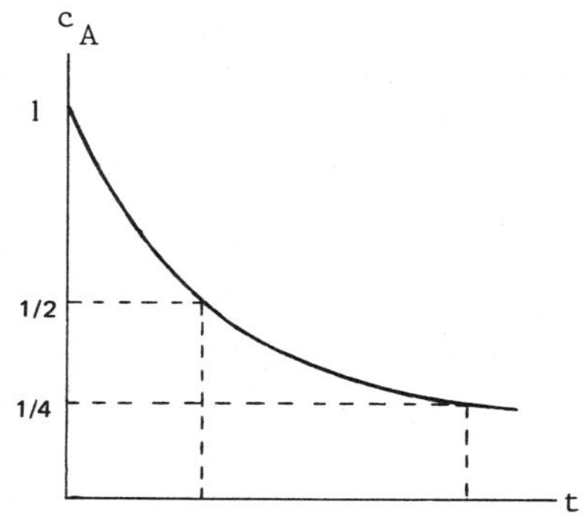

FIG. 2.4

At this point it would be of interest to consider various methods of ascertaining the reaction order. In other words we can inquire into the magnitude of n and its determination in the equation $-dc/dt = kc^n$. The solution is found most frequently by trial and error. For example, when n = 1, the plot of log c versus time must give a straight line. On the other hand, where n = 2, the plot of $1/c_A - 1/c_A^0$ versus time must give a straight line.

The reaction order can be determined more directly as follows: The experimentally determined concentrations are plotted against time (Fig. 2.5). If this graph is exactly known, it is possible to calculate the slope dc/dt at any point:

$$\frac{-dc}{dt} = kc^n$$

$$\log \frac{-dc}{dt} = \log k + n \log c$$

The order n equals the slope of the straight line which is formed by plotting log(-dc/dt) versus log c (Fig. 2.6).

Another method of determining the reaction order involves using the half-life. For a first-order reaction $t_{1/2}$ is independent of the initial concentration, while for a second-order reaction $t_{1/2}$ is inversely proportional to the reciprocal of the initial concentration. The general relation between reaction order and half-life can be derived as follows:

$$\frac{-dc}{dt} = kc^n \quad \text{or} \quad \frac{-dc}{c^n} = k\,dt$$

After integration and for $n \neq 1$,

$$\frac{1}{n-1} c^{1-n} = kt + \text{constant}$$

For $t = 0$ and $c = c^0$,

$$\frac{1}{n-1} c^{0^{1-n}} = \text{constant}$$

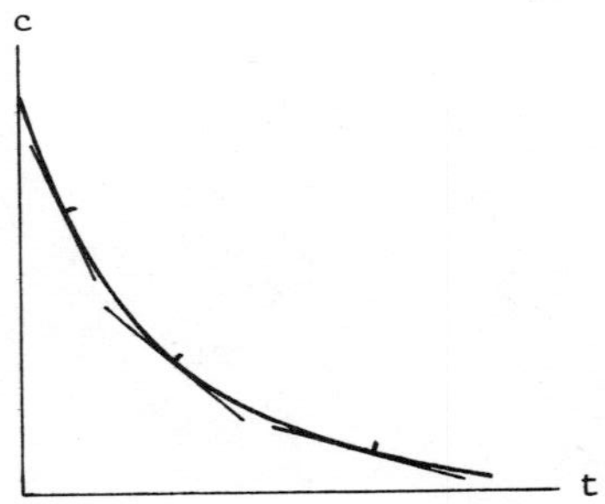

FIG. 2.5

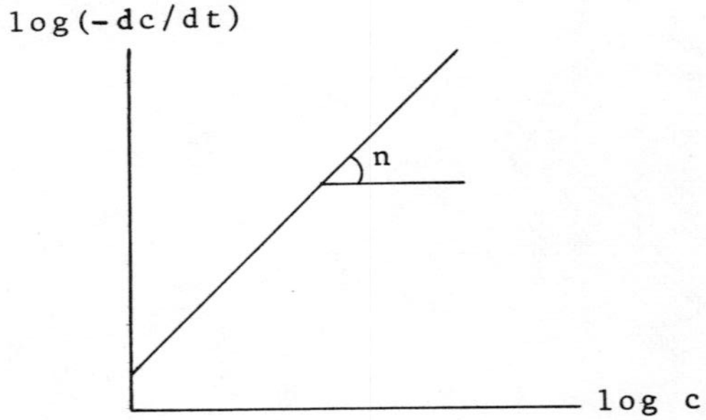

FIG. 2.6

$$\frac{1}{n - 1}\left(c^{1-n} - c^{0^{1-n}}\right) = kt$$

If $c = (1/2)c^0$,

$$\frac{c^{0^{1-n}}}{n - 1}\left(\frac{1}{2}^{1-n} - 1\right) = kt_{1/2}$$

and

$$t_{1/2} = \frac{2^{n-1} - 1}{k(n - 1)c^{0^{n-1}}} \qquad (2.13)$$

Thus, for a given reaction of the order n the length of a half-life period is inversely proportional to the initial concentration of that period raised to the (n - 1)st power.

For other methods of determining the order of a reaction, see C. H. Bamford and C. F. H. Tipper, eds., Comprehensive Chemical Kinetics, Vol. 1 (Elsevier, Amsterdam, 1969), p. 351.

Frequently, a reaction is first order in A and first order in B:

$$A + B \longrightarrow \text{products}$$

$$\frac{-dc_A}{dt} = \frac{-dc_B}{dt} = kc_Ac_B \qquad (2.14)$$

The reaction is first order in each component and, therefore, second order overall. The initial concentrations are c_A^0 and c_B^0:

$$c_A^0 - c_A = c_B^0 - c_B$$

Equation (2.14) can be solved by considering the general equation $dx/dt = -kxy$, which can be rewritten as

$$\frac{dx}{x} = -ky\,dt \qquad \text{or} \qquad d(\ln x) = -ky\,dt$$

Integration:

$$\int_0^t d(\ln x) = -k \int_0^t y\,dt$$

or

$$\ln \frac{x_t}{x_{t=0}} = -k \int_0^t y\,dt$$

Thus, from (2.14) we get

$$\ln \frac{c_A}{c_A^0} = -k \int_0^t c_B\,dt \qquad \text{and} \qquad \ln \frac{c_B}{c_B^0} = -k \int_0^t c_A\,dt$$

or

$$\ln \frac{c_A c_B^0}{c_A^0 c_B} = -k \int_0^t (c_B - c_A)\,dt$$

Since for this reaction $c_B - c_A = c_B^0 - c_A^0$, then

$$\ln \frac{c_A c_B^0}{c_A^0 c_B} = k(c_A^0 - c_B^0)t \tag{2.15}$$

Instead of c_A^0 and c_B^0, the symbols a and b are often used for the initial concentrations of the compounds A and B, while the decrease of the concentration is given by the symbol x: $c_A = a - x$ and $c_B = b - x$. With these symbols the equation becomes

$$\frac{1}{a - b} \ln \frac{b(a - x)}{a(b - x)} = kt \tag{2.16}$$

With the aid of Eq. (2.16) and data representing the variation of reactant concentration with time, it is possible to calculate the rate constant k. Another method of obtaining k is plotting $\ln[(a - x)/(b - x)]$ versus the time t. The slope of the straight line is equal to $k(a - b)$.

Equation (2.16) can hardly be used when a is nearly equal to b. In that case the logarithm is little different from zero. It is then advisable to alter the equation. Let

$$\frac{a + b}{2} = d \qquad \text{and} \qquad \frac{a - b}{2} = s$$

Then

$$a = d + s \qquad \text{and} \qquad b = d - s$$

Substitution into Eq. (2.16) gives

$$\frac{1}{2s}\left(\ln \frac{d - s}{d + s} + \ln \frac{d + s - x}{d - s - x}\right) = kt$$

$$\ln \frac{1 - s/d}{1 + s/d} + \ln \frac{1 + s/(d - x)}{1 - s/(d - x)} = 2kst$$

$$\ln\left(1 - \frac{s}{d}\right) - \ln\left(1 + \frac{s}{d}\right) + \ln\left(1 + \frac{s}{d - x}\right) - \ln\left(1 - \frac{s}{d - x}\right) = 2kst$$

The logarithm can be expressed in a series expansion as

$$\ln(1 + p) = p - \frac{1}{2} p^2 + \frac{1}{3} p^3 - \cdots$$

When $p \ll 1$ the higher terms can be neglected.

Applied to our equation

$$- \frac{s}{d} - \frac{1}{2}\frac{s^2}{d^2} - \frac{1}{3}\frac{s^3}{d^3} - \frac{s}{d} + \frac{1}{2}\frac{s^2}{d^2} - \frac{1}{3}\frac{s^3}{d^3} + \frac{s}{d - x} - \frac{1}{2}\frac{s^2}{(d - x)^2}$$

$$+ \frac{1}{3}\frac{s^3}{(d - x)^3} + \frac{s}{d - x} + \frac{1}{2}\frac{s^2}{(d - x)^2} + \frac{1}{3}\frac{s^3}{(d - x)^3} = 2kts$$

$$- \frac{2s}{d} - \frac{2}{3}\frac{s^3}{d^3} + \frac{2s}{d - x} + \frac{2}{3}\frac{s^3}{(d - x)^3} = 2kst$$

After division by 2s

$$\frac{1}{d - x} - \frac{1}{d} + \frac{s^2}{3}\left[\frac{1}{(d - x)^3} - \frac{1}{d^3}\right] = kt \qquad (2.17)$$

If a is equal to b, then $a = b = d$ and $s = 0$. The equation thus reduces to $1/(a - x) - 1/a = kt$, which has the same form as Eq. (2.11) derived earlier from $-dc_A/dt = kc_A^2$.

The dimensions of k follow from the equation; they are $1/(\text{time} \times \text{concentration})$. The usual units are liters $\text{mol}^{-1}\ s^{-1}$.

$$1\ \text{liter mol}^{-1}\ \text{min}^{-1} = \frac{1}{60}\ \text{liter mol}^{-1}\ s^{-1}$$

BROKEN ORDERS

Many rates can be analyzed with either first- or second-order rate equations, but many more rate equations are possible. For example, a reaction can be of third order or can involve a fractional order. A fractional order often indicates dissociation. For example, a compound S decomposes under the catalytic influence of a carboxylic acid:

$$RCOOH \rightleftharpoons RCOO^- + H^+ \qquad \text{rapid}$$

$$H^+ + S \longrightarrow Y \qquad \text{slow}$$

$$Y \longrightarrow Z + H^+ \qquad \text{rapid}$$

From the rate-determining step we conclude that $v = kc_{H^+} c_S$. In accordance with the rapid equilibrium,

$$\frac{c_{RCOO^-} c_{H^+}}{c_{RCOOH}} = K$$

Since $c_{RCOO^-} = c_{H^+}$ (if no additional $RCOO^-$ or H^+ is added), $(c_{H^+})^2 = Kc_{RCOOH}$. If the dissociation equilibrium of the carboxylic acid lies far to the left, c_{RCOOH} approximately corresponds to the added amount of carboxylic acid:

$$c_{H^+} = K^{1/2} c_{RCOOH}^{1/2}$$

Thus

$$v = kK^{1/2} c_{RCOOH}^{1/2} c_S = k' c_{RCOOH}^{1/2} c_S \qquad (2.18)$$

See the tables in C. H. Bamford and C. F. H. Tipper, eds., Comprehensive Chemical Kinetics, Vol. 1 (Elsevier, Amsterdam, 1969), p. 361, and S. W. Benson, The Foundations of Chemical Kinetics (McGraw-Hill, New York, 1960), p. 24.

In this case the dimensions of k' are $1/(c^{1/2}t)$; its preferred units are $\text{liters}^{1/2}\ \text{mol}^{-1/2}\ \text{s}^{-1}$.

An example of a very intricate rate equation has been observed for the gas phase reaction

$$H_2 + Br_2 \rightleftarrows 2HBr$$

Hydrogen and bromine react in the dark at temperatures between 200 and 300°C. The rate of formation of HBr is given by

$$\frac{dc_{HBr}}{dt} = \frac{kc_{H_2}c_{Br_2}^{1/2}}{1 + k'c_{HBr}/c_{Br_2}} \tag{2.19}$$

$k' \simeq 1/10$ and is independent of temperature.

At the beginning of reaction, when $c_{HBr} \ll c_{Br_2}$, the equation reduces to $dc_{HBr}/dt = kc_{H_2}c_{Br_2}^{1/2}$.

The half order in bromine stems from the fact that this reaction is initiated by the dissociation

$$Br_2 \underset{}{\overset{K}{\rightleftarrows}} 2Br$$

This equilibrium, lying far to the left, can be expressed in the form

$$c_{Br} = K^{1/2}c_{Br_2}^{1/2}$$

It must be emphatically stated that the order of a reaction cannot be simply derived from the stoichiometric equation. As we have seen earlier, the stoichiometry of a reaction between A and B to give product may result in the rate equation

$$v = kc_A c_B$$

But this is not a necessary consequence. For example, let A be a keto compound which slowly converts to its enol isomer. If the enol reacts with B in a rapid step, the concentration of B does not appear in the rate equation. A similar situation occurs when the first step is a slow isomerization of A from a trans to a very reactive cis configuration or from ground state to excited state.

PSEUDO-ORDERS

Sometimes the expressions *pseudo-first order* or *pseudo-second order* are used. The following example is a case in point:

$$RCOOR' + H_2O \longrightarrow RCOOH + R'OH$$

The rate of this ester hydrolysis depends on the ester and water concentrations, and consequently, the reaction process is second order. However, if the hydrolysis takes place in an aqueous environment, the water concentration is large and practically invariant. Experimentally it is found that

$$v = kc_{RCOOR'}$$

The reaction is designated as pseudo-first order because of all concentrations which appear in the complete rate law only one is varying measurably. Therefore, in the experimental rate equation the non-varying concentrations have been included in the "apparent" rate constant k.

Another example:

$$(CH_3)_3CN{\equiv}C + H_2O \longrightarrow (CH_3)_3C{-}\overset{H}{\overset{|}{N}}{-}\overset{H}{\overset{|}{C}}{=}O$$

The addition of water to tert-butyl isocyanide is catalyzed by acids. If water is used as a solvent containing a small amount of acid, for example, perchloric acid, the rate equation is found to be

$$\frac{-dc}{dt} = kc_{isocyanide}$$

Apparently, we have a first-order reaction, and by using 0.01 mol/liter of $HClO_4$ at 25°C, $k = 4.49 \times 10^{-4}\ s^{-1}$. Under the same conditions but with 0.02 mol/liter of $HClO_4$, $k = 9.01 \times 10^{-4}\ s^{-1}$. Consequently, the pseudo-first-order rate constant k appears to be directly proportional to the acid concentration. A more complete rate equation is, therefore,

W. Drenth, Recl. Trav. Chim. Pays-Bas <u>81</u>:319 (1962).

$$\frac{-dc}{dt} = k'c_{H_3O^+}c_{isocyanide}$$

in which the bimolecular rate constant $k' = 0.045$ liter $mol^{-1}\ s^{-1}$. Notice should be taken of the difference in the units of k and k'. The reaction is clearly second order, but in the course of a given run we observe only pseudo-first-order behavior because the concentration of the hydronium ions remains constant.

For the oxidation of acetoin by ferricenium ions in acidified, aqueous alcohol pseudo-zero-order behavior has been observed:

$$H_3C-\overset{\underset{|}{OH}}{\underset{H}{C}}-\overset{\overset{O}{\|}}{C}-CH_3 + 2(C_5H_5)_2Fe^+ \longrightarrow H_3C-\overset{\overset{O}{\|}}{C}-\overset{\overset{O}{\|}}{C}-CH_3 + 2(C_5H_5)_2Fe + 2H^+$$

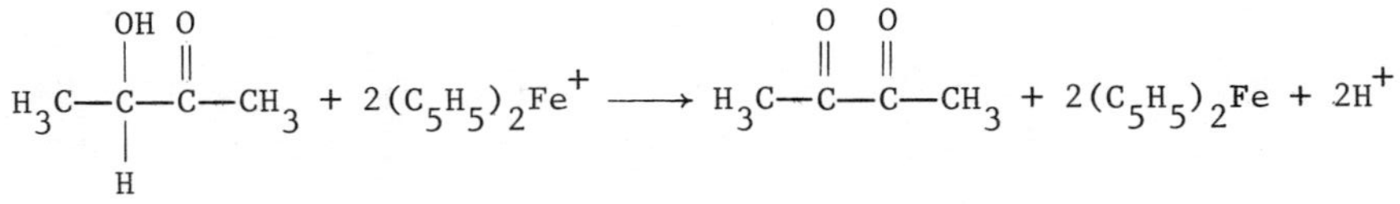

The reaction was performed with an excess of acetoin. The rate was found to be independent of the ferricenium concentration as long as this concentration is not very small (see Fig. 2.7). The rate-determining step is the enolization of the acetoin:

$$CH_3-\overset{\underset{|}{OH}}{\underset{H}{C}}-\overset{\overset{O}{\|}}{C}-CH_3 \xrightarrow{H^+} CH_3-\overset{\underset{|}{OH}}{C}=\overset{\underset{|}{OH}}{C}-CH_3$$

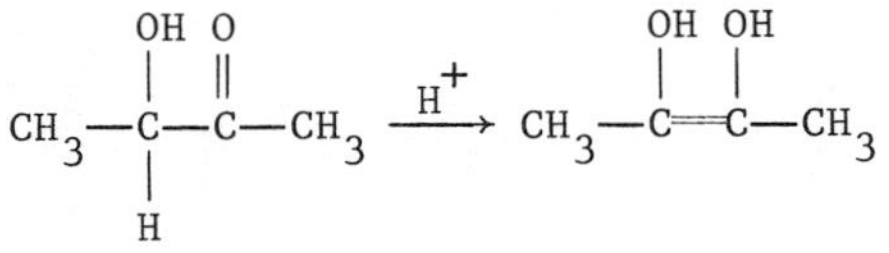

Rapid oxidation of the enol follows. Because of the excess of acetoin, the rate of formation of the enol and, therefore, the rate of decrease of concentration of the ferricenium ions are practically constant.

These examples illustrate the fact that the reaction kinetics can be simplified through maintaining a nearly constant concentration of one reactant:

$$v = kc_A^{\alpha} c_B^{\beta}$$

The order α of A can be determined easily by keeping the concentration of B constant and likewise with regard to the order β of B.

[Ferricenium]

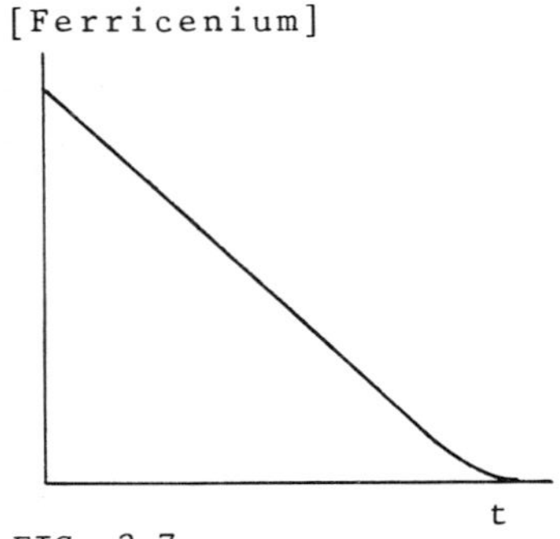

FIG. 2.7

J. Lubach and W. Drenth, Recl. Trav. Chim. Pays-Bas 89:144 (1970).

How does one keep a concentration constant? One way is to use a large excess of a reactant, e.g., 10 to 50 times the equivalent amount. With protons or hydroxyl ions the pH can be kept constant with an automatic titrator, a pH-stat. Sometimes it is possible to keep the concentration constant by using a saturated solution. However, the rate of solution under these circumstances has to be faster than the reaction rate. Commonly, in such cases, vigorous stirring will be required (Fig. 2.8).

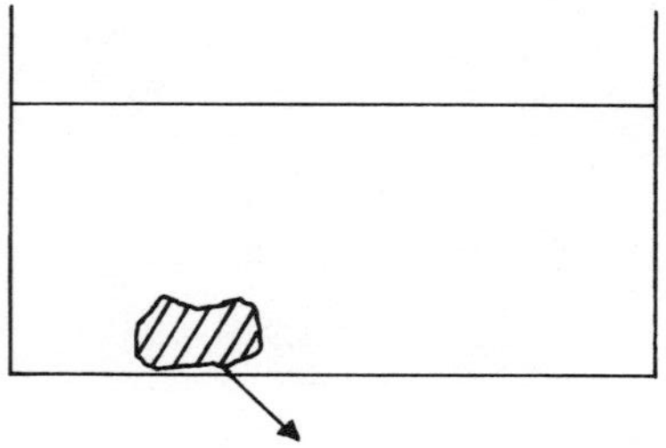

Neat compound solid or liquid

FIG. 2.8

COMPETING AND CONSECUTIVE REACTIONS

The electrophilic bromination of toluene results in a parallel formation of ortho-, meta-, and para-bromotoluene:

$$\text{Toluene} + Br_2 \rightarrow \begin{cases} \text{o-Br-toluene} & v_o = 2k_o c_{tol} c_{Br_2} \\ \text{m-Br-toluene} & v_m = 2k_m c_{tol} c_{Br_2} \\ \text{p-Br-toluene} & v_p = k_p c_{tol} c_{Br_2} \end{cases}$$

If the three compounds can be assumed to be noninterconvertible under the reaction conditions and if consecutive reactions can be neglected, then

$$2k_o : 2k_m : k_p = o:m:p$$

where o, m, and p are the respective concentrations of ortho, meta, and para products. The relative rates follow directly from the product

L. M. Stock and H. C. Brown, Advan. Phys. Org. Chem. 1:35 (1963).

distribution. H. C. Brown used this approach in his studies of selectivity in electrophilic aromatic substitutions. The following product distributions were found under the specified conditions:

	o (%)	m (%)	p (%)
Noncatalytic bromination	32.9	0.3	66.8
Isopropylation with i-PrCl + $AlCl_3$	46.9	17.4	35.7
When nonselective (randomly)	40	40	20

The selectivity is much higher in this bromination than in this alkylation reaction.

If the total rate k_{tot} is known, for example, from measurements of the bromine concentration, then the partial rate constants can be calculated as follows:

$$k_o = \frac{(1/2)o}{o + m + p} k_{tot}$$

$$k_m = \frac{(1/2)m}{o + m + p} k_{tot}$$

$$k_p = \frac{p}{o + m + p} k_{tot}$$

This approach to calculating partial rate factors is, however, not always permissible. Take, for example, the solvolysis of tert-butyl chloride in 60 vol % alcohol at 25°C. The product composition is found to be 17% isobutylene and 83% (tert-butyl alcohol′ + tert-butyl ethyl ether):

C. K. Ingold, Structure and Mechanism in Organic Chemistry (Cornell Univ. Press, Ithaca, N.Y., 1953), p. 426.

$$(CH_3)_3CCl \xrightarrow{k_1} \begin{array}{l} (CH_3)_2C = CH_2 \\ \\ (CH_3)_3COR \quad R = H \text{ or } C_2H_5 \end{array}$$

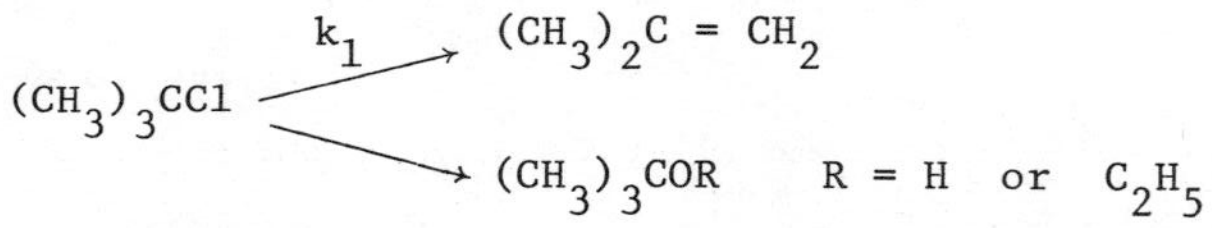

One cannot conclude that $k_1 = 0.17 \times k_{tot}$, since the reaction mechanism is much more complicated. The products are formed from a common intermediate:

$$(CH_3)_3CCl \xrightarrow[(1)]{slow} (CH_3)_3C^+ + Cl^- \begin{array}{l} \swarrow (2) \quad \downarrow (3) \quad \searrow (4) \\ \text{isobutylene alcohol ether} \end{array}$$

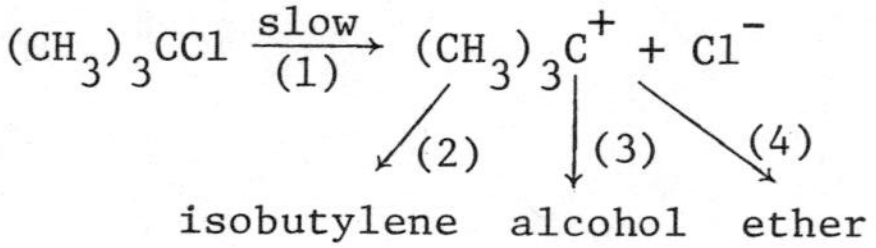

Although $v_2 : v_3 : v_4 =$ amount of isobutylene:amount of alcohol:amount of ether, the partial rate factors of steps 2, 3, and 4 cannot be calculated from this experiment. The overall rate is controlled entirely by the first step in which the common intermediate is formed. The competition involves only product-forming steps which are much faster than the rate-determining step, and their rates bear no relation to the rate of formation of the common intermediate.

Another example of consecutive reactions is the halogenation of acetone, which has been discussed earlier (page 6). Reactions with a large number of consecutive steps are the chain reactions such as the formation of HBr from H_2 and Br_2 in the dark at 300°C in the gas phase. This reaction has been mentioned on page 26 to illustrate an intricate rate equation. The experimental data conform to the rate law

$$\frac{d[HBr]}{dt} = \frac{k[H_2][Br_2]^{1/2}}{1 + k'[HBr]/[Br_2]} \tag{2.20}$$

To ascertain the reaction mechanism in such a case the following approach is used. Many different conceivable mechanisms are written, and for each of them the rate equation is derived. The mechanism which leads to an equation equivalent to the observed rate equation will receive further consideration. In the present example this mechanism is

(1) $Br_2 \xrightarrow{k_1} 2Br$

(2) $Br + H_2 \xrightarrow{k_2} HBr + H$

(3) $H + Br_2 \xrightarrow{k_3} HBr + Br$

(4) $H + HBr \xrightarrow{k_4} H_2 + Br$

(5) $2Br \xrightarrow{k_5} Br_2$

The first step is the initiation step. Steps 2, 3, and 4 are the propagation steps. The last step, which consumes radicals without forming new ones, is the termination step. In this case the last step is also called a recombination step because the radicals dimerize. The rate of formation of HBr is

$$\frac{d[HBr]}{dt} = k_2[Br][H_2] + k_3[H][Br_2] - k_4[H][HBr] \tag{2.21}$$

In complex reaction processes which involve formation of reactive intermediates I in low concentration, it is possible to introduce a simplifying assumption called the steady-state approximation. This applies only when it can be justified that during the main reaction period the variation of the given reaction intermediate concentration is quite small; i.e., $dc_I/dt \simeq 0$. In the present instance there are two identifiable intermediates of this nature, hydrogen and bromine atoms. Their concentrations become relatively invariant after a brief induction period of building up the steady state and prior to the tail end of reaction when their concentrations are declining from the steady-state values:

$$\frac{d[H]}{dt} = k_2[Br][H_2] - k_3[H][Br_2] - k_4[H][HBr] = 0 \qquad (2.22)$$

$$\frac{d[Br]}{dt} = 2k_1[Br_2] - k_2[Br][H_2] + k_3[H][Br_2] + k_4[H][HBr] - 2k_5[Br]^2 = 0 \qquad (2.23)$$

Subtracting (2.22) from (2.21) gives

$$\frac{d[HBr]}{dt} = 2k_3[H][Br_2] \qquad (2.24)$$

Adding (2.22) to (2.23) gives $2k_1[Br_2] - 2k_5[Br]^2 = 0$; therefore,

$$[Br] = \left(\frac{k_1}{k_5}\right)^{1/2} [Br_2]^{1/2} \qquad (2.25)$$

Insertion of (2.25) into (2.22) gives

$$k_2\left(\frac{k_1}{k_5}\right)^{1/2}[Br_2]^{1/2}[H_2] - (k_3[Br_2] + k_4[HBr])[H] = 0$$

$$[H] = \frac{k_2(k_1/k_5)^{1/2}[H_2][Br_2]^{1/2}}{k_3[Br_2] + k_4[HBr]} \tag{2.26}$$

Insertion of (2.26) into (2.24) gives

$$\frac{d[HBr]}{dt} = \frac{2k_3k_2(k_1/k_5)^{1/2}[H_2][Br_2]^{3/2}}{k_3[Br_2] + k_4[HBr]}$$

Division by $k_3[Br_2]$ gives

$$\frac{d[HBr]}{dt} = \frac{2k_2(k_1/k_5)^{1/2}[H_2][Br_2]^{1/2}}{1 + (k_4/k_3)([HBr]/[Br_2])}$$

This equation agrees with all the experimentally determined rate dependencies. Therefore, from a kinetic viewpoint it is justified to accept the proposed mechanism.

Rate versus concentration curves as in Fig. 2.9 are often observed in enzyme catalyzed reactions. This kinetic behavior is explained by assuming the intermediate formation of a complex (ES) from enzyme (E) and substrate (S):

$$E + S \underset{k_{-1}}{\overset{k_1}{\rightleftharpoons}} ES \qquad ES \xrightarrow{k_2} \text{products} + E$$

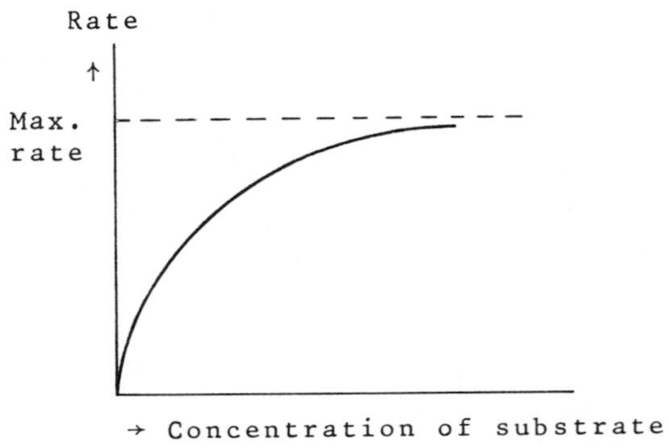

FIG. 2.9

The rate of product formation depends on the concentration of ES,

$$v = k_2[ES] \tag{2.27}$$

When the complex is formed in a rapid equilibrium

$$\frac{[E][S]}{[ES]} = \frac{k_{-1}}{k_1} = K_e \tag{2.28}$$

$$[E] = [E_t] - [ES] \tag{2.29}$$

where $[E_t]$ is the total enzyme concentration; we assume $[S] >> [ES]$.

From (2.28) and (2.29)

$$[ES] = \frac{[E_t][S]}{[S] + K_e} \tag{2.30}$$

and

$$v = \frac{k_2[E_t][S]}{[S] + K_e} \tag{2.31}$$

Equation (2.31) is called the Michaelis-Menten equation.

A similar equation is obtained when the complex ES is not in rapid equilibrium with E and S. Since, generally, ES is present in low concentration, we can apply the steady-state treatment:

$$\frac{d[ES]}{dt} = k_1[E][S] - k_{-1}[ES] - k_2[ES] = 0 \tag{2.32}$$

From (2.29) and (2.32) it follows that

$$[ES] = \frac{k_1}{k_1[S] + k_{-1} + k_2}[E_t][S] \qquad (2.33)$$

and

$$v = \frac{k_1 k_2}{k_1[S] + k_{-1} + k_2}[E_t][S] = \frac{k_2}{[S] + K_m}[E_t][S] \qquad (2.34)$$

where $K_m = (k_{-1} + k_2)/k_1$ is called the Michaelis constant.

Equations (2.34) and (2.31) are similar in form. At high substrate concentration $[S] >> K_m$ and $v_{max} = k_2[E_t]$. Equation (2.34) can also be written as

$$v = \frac{v_{max}}{[S] + K_m}[S] \qquad (2.35)$$

The inverted form of this equation,

$$\frac{1}{v} = \frac{1}{v_{max}} + \frac{K_m}{v_{max}[S]} \qquad (2.36)$$

shows that a plot of $1/v$ versus $1/[S]$ is linear. Such a plot is called a Lineweaver-Burk plot. From its intercept and slope v_{max} and K_m can easily be calculated.

H. Lineweaver and D. Burk, J. Amer. Chem. Soc. 56:658 (1934).

In practice it is found that many organic reactions are attended by side reactions occurring in parallel or in succession to the reaction of particular interest. While most situations of this degree of complexity cannot be analyzed in a straightforward manner, they are nonetheless susceptible to a kinetic treatment which can best be discussed with the aid of the typical system

$$A \xrightarrow{k_1} B, \quad B \xrightarrow{k_2} C, \quad B \xrightarrow{k_3} D$$

In such cases the use of computers is required to calculate the various rate constants. As an example we mention the industrial production of pentaerythritol, $(HOCH_2)_4C$, from formaldehyde and pentaerythrose in basic solution. The pentaerythrose is obtained in an aldol condensation reaction from formaldehyde and acetaldehyde.

M. S. Peters and C. R. Cupit, Chem. Eng. Sci. 10:57 (1959).

$$H_2CO + CH_3CHO \rightarrow HOCH_2CH_2CHO$$
$$H_2CO + HOCH_2CH_2CHO \rightarrow (HOCH_2)_2CHCHO$$
$$H_2CO + (HOCH_2)_2CHCHO \rightarrow (HOCH_2)_3CCHO$$

Two Cannizzaro-type reactions occur; the main reaction is

$$H_2CO + (HOCH_2)_3CCHO + OH^- \xrightarrow{k_1} (HOCH_2)_4C + HCOO^-$$

and a side reaction is

$$2H_2CO + OH^- \xrightarrow{k_2} H_3COH + HCOO^-$$

We have need to determine the rate constants k_1 and k_2. The reactions are of the type

$$A + B + C \xrightarrow{k_1} \text{main products}$$
$$2A + C \xrightarrow{k_2} \text{side products}$$

D. M. Himmelbau, C. R. Jones, and K. B. Bischoff, Ind. Eng. Chem. Fund. 6:539 (1967). A FORTRAN code for the CDC 6600 digital computer is available from the authors.

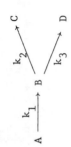

The corresponding rate equations are

$$\frac{d[A]}{dt} = -k_1[A][B][C] - k_2[A]^2[C] \tag{2.37}$$

$$\frac{d[B]}{dt} = -k_1[A][B][C] \tag{2.38}$$

$$\frac{d[C]}{dt} = -k_1[A][B][C] - \frac{k_2}{2}[A]^2[C] \tag{2.39}$$

Each of these equations is integrated from $t = t_0$ to $t = t_i$:

$$[A]_{t_i} - [A]_{t_0} = -k_1 \int_{t_0}^{t_i} [A][B][C]\ dt - k_2 \int_{t_0}^{t_i} [A]^2[C]\ dt \tag{2.40}$$

$$[B]_{t_i} - [B]_{t_0} = -k_1 \int_{t_0}^{t_i} [A][B][C]\ dt \tag{2.41}$$

$$[C]_{t_i} - [C]_{t_0} = -k_1 \int_{t_0}^{t_i} [A][B][C]\ dt - \frac{k_2}{2} \int_{t_0}^{t_i} [A]^2[C]\ dt \tag{2.42}$$

At various times t_i the concentrations [A], [B], and [C] are estimated by some analytical procedure of sufficient accuracy and precision. Thus, for each of the compounds A, B, and C the concentration versus time relation is known (Fig. 2.10). These relations permit, through use of the computer, a numerical computation of the integrals such as

$$\int_{t_0}^{t_i} [A][B][C]\ dt$$

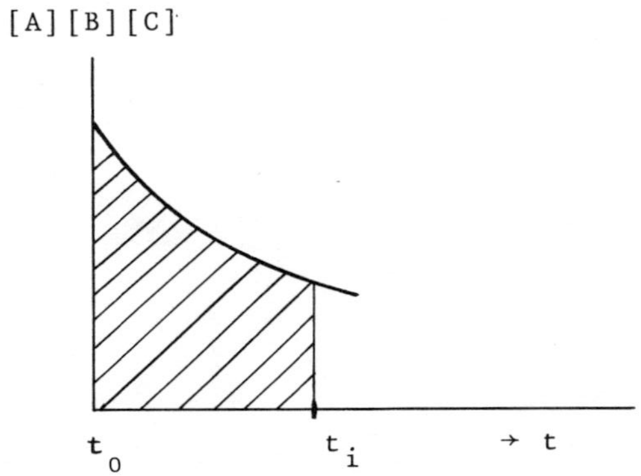

FIG. 2.10

Thus, for each time t_i three equations [(2.40), (2.41), and (2.42)] are obtained which are linear in k_1 and k_2. From these equations k_1 and k_2 are computed by a least-squares method, minimizing

$$\sum_i [([A]_{t_i} - [A]_{t_0})_{calc} - ([A]_{t_i} - [A]_{t_0})_{obs}]^2 \qquad (2.43)$$

and the analogous forms for [B] and [C]. It is advisable to include in this calculation data gathered in a series of runs if the most precise values of the rate constants are to be realized. The rate constants of the pentaerythritol process determined in this way from data at approximately 30°C are

$$k_1 = 1.95 \times 10^{-3} \text{ liter}^2 \text{ mol}^{-2} \text{ s}^{-1}$$

and

$$k_2 = 0.0165 \times 10^{-3} \text{ liter}^2 \text{ mol}^{-2} \text{ s}^{-1}$$

3 Experimental Procedures

Tools presently at hand

In general, it is sufficient to follow the changes in concentrations of one of the participating compounds. A variety of analytical methods are often available. A number of examples will be given without any pretention of being complete. A simple method, for example, is direct titration, which will be illustrated with the solvolysis of isopropyl bromide in 70 vol % aqueous alcohol at 75°C. In an actual run 0.2801 g of isopropyl bromide was dissolved in the solvent and diluted to 100 cm^3. Using a syringe, samples of this solution were transferred and sealed into each of 10 ampules. The ampules were placed in a constant temperature bath maintained at 75°C. Care was taken to assure that the free volume in the ampules was small. At definite intervals an ampule was opened and 5 cm^3 of the reacted sample titrated with sodium hydroxide.

$$(CH_3)_2CHBr + ROH \longrightarrow \left.\begin{matrix} (CH_3)_2CHOH \\ (CH_3)_2CHOC_2H_5 \\ CH_3CH{=}CH_2 \end{matrix}\right\} + HBr$$

In this fashion the concentration of HBr which had formed could be computed. The concentration of HBr is equal to the decrease of the isopropyl bromide concentration. $c^0_{i\text{-PrBr}} = 0.02277$ mol/liter. See Table 3.1.

TABLE 3.1

Time (min)	NaOH (cm^3)	c_{HBr} (mol/liter)	$c_{i\text{-PrBr}}$ (mol/liter)	$\log \left(\frac{c_{i\text{-PrBr}}}{\text{mol/liter}} \right)$	$\log \frac{c^0_{i\text{-PrBr}}}{c_{i\text{-PrBr}}}$	0.4343k (min^{-1}) ($\times 10^4$)
0	0	0	0.02277	0.3574 - 2	0	
64	0.277	0.00255	0.02022	0.3058 - 2	0.0516	8.06
103	0.438	0.00403	0.01874	0.2728 - 2	0.0846	8.21
169	0.670	0.00617	0.01660	0.2201 - 2	0.1373	8.12
262	0.956	0.00880	0.01397	0.1452 - 2	0.2122	8.10
326	1.126	0.01037	0.01240	0.0934 - 2	0.2640	8.10
492	1.486	0.01368	0.00909	0.9586 - 3	0.3988	8.11
611	1.681	0.01548	0.00729	0.8627 - 3	0.4947	8.10
692	1.799	0.01656	0.00621	0.7931 - 3	0.5643	8.15

Average: $(8.12 \pm 0.02) \times 10^{-4}$ min^{-1}

$$k = \frac{8.12 \times 10^{-4}}{0.434 \times 60} = (3.12 \pm 0.01) \times 10^{-5}\ s^{-1}$$

FIG. 3.1

The constants a and b in the equation $y = ax + b$ for n pairs of values of y and x are given by

$$a = \frac{n\Sigma xy - \Sigma x \Sigma y}{n\Sigma x^2 - (\Sigma x)^2} \qquad b = \frac{\Sigma y - a\Sigma x}{n}$$

The equation for a first-order reaction was applied:

$$\log c = \log c^0 - 0.434kt$$

Instead of calculating k for each point separately, it is possible to plot log c versus the time. The slope of the straight line obtained can be calculated from the plot or by a least-squares treatment. The latter leads to $k = (3.12 \pm 0.01) \times 10^{-5}\ s^{-1}$.

For calculation of the standard deviation and regression coefficient, see, e.g., J. Shorter, Correlation Analysis in Organic Chemistry (Clarendon Press, Oxford, 1973), p. 103. References to computer programs for chemical kinetics have been compiled by J. L. Hogg, J. Chem. Educ. 51:109 (1974).

In the isopropyl bromide solvolysis example the k-values are reasonably constant during a run. Sometimes it is found that the k-values are not constant. For instance, in the solvolysis of dilute p,p'-dimethylbenzhydryl chloride in 90% aqueous acetone at 0°C the first-order rate shows a considerable degree of nonrandom inconstancy during the course of reaction.

C. K. Ingold, Structure and Mechanism in Organic Chemistry (Cornell Univ. Press, Ithaca, N.Y., 1953), p. 364.

$$(CH_3C_6H_4)_2CHCl + H_2O \longrightarrow (CH_3C_6H_4)_2CHOH + HCl$$

conversion (%)	$k_{relative}$
0	1
20	0.90
40	0.82
60	0.78
80	0.74

These rate constant values were computed on the basis of the equation for a first-order reaction. The observation of a unidirectional drift in k clearly signifies that the reaction is not first order.

This phenomenon can be interpreted as a special case of a solvolytic process of the type S_N1. The mechanism consists of the following steps:

(1)
(2)
(3)

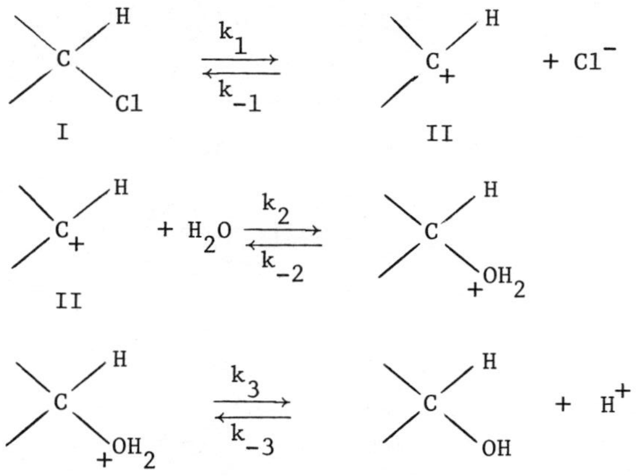

Usually, for reactions of the type S_N1, step 1 is rate determining and steps 2 and 3 are fast, consecutive reactions. Under these circumstances

$$v = k_1 \times c_{\text{benzhydryl chloride}}$$

However, if the rate constant k_2 approaches the value of k_{-1}, then the situation is more complicated. In deriving the appropriate rate equation the steady-state approximation must be applied. In this approximation the variation in concentration of the reactive carbocation with time $dc_{II}/dt = 0$, or

$$k_1c_I - (k_{-1}c_{II}c_{Cl^-}) - k_2c_{II} = 0$$

where k_2 includes the concentration of the H_2O;

$$c_{II} = \frac{k_1 c_I}{k_2 + k_{-1} c_{Cl^-}}$$

Step 2 can be neglected because step 3, the ionization of an OH bond, is much faster. The net reaction rate of species I is equal to the reaction rate for product formation from the carbocation and is thus given by

$$v = k_2 c_{II} = \frac{k_2 k_1 c_I}{k_2 + k_{-1} c_{Cl^-}} = \frac{k_1 c_I}{1 + (k_{-1}/k_2) c_{Cl^-}}$$

In fact an equation of this form is applicable to all S_N1 reaction processes. If, as is often the case, $k_2 >> k_{-1} c_{Cl^-}$, then the equation reduces to

$$v = k_1 c_I$$

and the reaction is recognized as a simple first-order process under the specified reaction conditions.

The initial chloride ion concentration is zero. During the reaction the chloride ion concentration increases, and the term $k_{-1} c_{Cl^-}$ increases along with it. Sometimes, e.g., in the solvolysis of p,p'-dimethylbenzhydryl chloride, $k_{-1} c_{Cl^-}$ attains a magnitude which is comparable to that of k_2. The denominator thus becomes greater, and the reaction rate decreases. This decrease of rate can also be affected by the addition of chloride ion salts.

This phenomenon is called the mass law effect by analogy to the mass law effect in thermodynamics, which reflects the equilibrium

$$A \rightleftarrows B + C$$

shifting to the left with increasing concentration of C.

An example of a second-order reaction using titration as a method of analysis is the hydrolysis of $(CH_3)_3CCH_2CH_2COOC_2H_5$ in 84.8 weight % aqueous ethyl alcohol containing NaOH at 25.0°C. The ester and the NaOH were dissolved separately. The two solutions were mixed and kept in a constant-temperature bath. The moment of mixing the ester and NaOH solutions is designated as t = 0. The initial concentrations of the ester, a, and of the NaOH, b, are both 0.0474 mol/liter. At fixed intervals 10 cm^3 was pipetted from the reaction mixture and added to an excess of hydrogen chloride solution. The reaction stopped immediately. The HCl neutralized the OH^- ions as well as the $(CH_3)_3CCH_2CH_2COO^-$ ions. The excess of HCl and the acid $(CH_3)_3CCH_2CH_2COOH$ (concentration x) were now titrated with a standard base. See Table 3.2. The equation

$$(CH_3)_3CCH_2CH_2COOC_2H_5 + OH^- \longrightarrow (CH_3)_3CCH_2CH_2COO^- + C_2H_5OH$$

$$\frac{1}{a - x} - \frac{1}{a} = kt$$

is used for the calculation of each k-value. In this way the concentration a has excessive weight in calculations of the average k-value. This is avoided by plotting 1/(a - x) versus the time. A straight line is obtained. Least-squares calculation gives a slope equal to $k = (1.80 \pm 0.01) \times 10^{-3}$ liter mol^{-1} s^{-1}.

TABLE 3.2

Time (s)	x (mol/liter)	a - x = b - x (mol/liter)	1/(a - x) (liters/mol)	1/(a - x) - (1/a) (liters/mol)	k × 10^3 liters mol^{-1} s^{-1}
1,255	0.0043	0.0431	23.20	2.10	1.67
2,835	0.0089	0.0385	25.97	4.87	1.72
4,890	0.0138	0.0336	29.76	8.66	1.77
7,865	0.0190	0.0285	35.09	13.99	1.78
12,230	0.0241	0.0233	42.92	21.82	1.78
15,360	0.0267	0.0207	48.31	27.21	1.77
18,900	0.0293	0.0181	55.25	34.15	1.81

Average: $k = (1.76 \pm 0.04) \times 10^{-3}$ liters mol^{-1} s^{-1}

TITRATION

Titration can be performed automatically. The addition of reagent is controlled by the voltage between a suitable electrode and a reference electrode. An example is the use of such a "pH-stat" in the hydrolysis of p-nitrophenyl benzoate catalyzed by a series of substituted benzohydroxamic acid anions as nucleophiles:

H. Kwart and H. Omura, J. Org. Chem. 34:318 (1969).

$$C_6H_5COO{-}C_6H_4{-}p\text{-}NO_2 + H_2O \xrightarrow{p\text{-}X\text{-}C_6H_4CONHO^-} C_6H_5COOH + HOC_6H_4{-}p\text{-}NO_2$$

A glass electrode and a reference electrode were inserted in the reaction solution. The medium was brought to a pH of 11.0 and constantly maintained at this value by neutralizing the acidic products through addition of KOH solution. A pseudo-first-order rate constant k_1 was calculated from the data obtained by recording the amount of standard KOH solution required as a function of time to maintain pH = 11.0. This first-order rate constant obeys the equation

$$k_1 = k_{H_2O}c_{H_2O} + k_{OH^-}c_{OH^-} + k_{\text{benzohy. anion}}c_{\text{benzohy. anion}}$$

In the p-X-benzohydroxamic acid anion electron-donating substituents X increase the nucleophilicity and therefore the rate, whereas the reverse is true for electron-attracting substituents. This constitutes an example of nucleophilic catalysis of solvolytic reactivity.

X	$k_{\text{benzohy. anion}} \times 10^4$ (liters mol^{-1} s^{-1})
CH_3O	1.73
H	1.40
Cl	0.90
NO_2	0.22
$^+N(CH_3)_3I^-$	0.23

ULTRAVIOLET SPECTROMETRY

Other methods which have been used are gravimetric analysis; ultraviolet, infrared, proton magnetic resonance spectroscopy; and gas chromatography. Ultraviolet spectrophotometry is particularly useful for pursuit of reactions in systems involving compounds strongly absorbing in the near UV or visible regions. The advantage is that quantitative measurements can be performed accurately. Solvents such as water, alcohol, and cyclohexane do not absorb significantly. Moreover, very small concentrations, of the order of 10^{-3} mol/liter, can be pursued by this technique.

In the following example an ultraviolet spectrophotometer was used for the study of chemical reactivity in the Meyer-Schuster rearrangement of 1-ethylthio-3-hydroxy-3-methyl-1-butyne. The product of the reaction the α,β-unsaturated thiol ester, has an absorption maximum at λ_{max} = 267 nm with a molar absorption coefficient $\varepsilon_{max} = 1010\ m^2\ mol^{-1}$. At this wavelength the starting compound hardly absorbs. The purpose of the study was to look for the influence of acids on the ease of rearrangement. The starting compound was soluble in water. The solubility is not large but sufficient. What is the required concentration? Suppose that after total conversion the absorbance of the solution, A, is 0.8. According to the Lambert-Beer law, A = εcd. For a cell thickness d of 1 cm, the concentration is $c = 0.8/[(1010\ m^2\ mol^{-1})(1\ cm)] \simeq 0.08\ mol/m^3 = 0.8 \times 10^{-4}$ mol/liter.

The absorbance A of the solution was measured as a function of time. From A it was possible to calculate the concentration of the ester formed. Through application of the relation

$$c_{ester} = c^{\infty}_{ester}[1 - \exp(-kt)]$$

the rate constant k was readily computed. At a temperature of 25°C and

G. L. Hekkert and W. Drenth, Recl. Trav. Chim. Pays-Bas 80:1285 (1961).

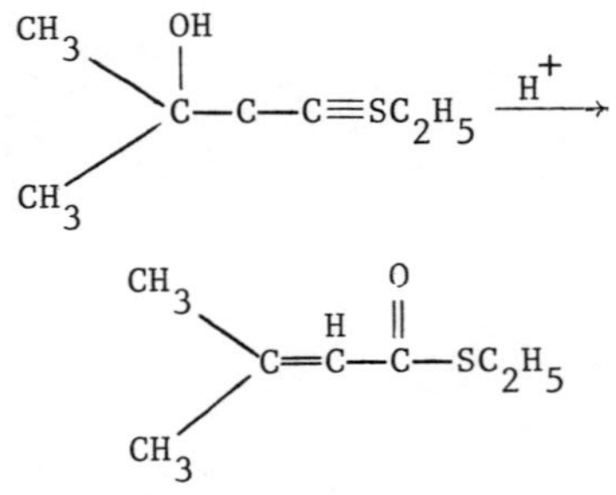

The traditional units for the molar absorption coefficient are liters/mol cm = 0.1 m^2/mol.

using water as a solvent, the results were

$$k = 11.6 \times 10^{-5}\ s^{-1} \qquad \text{at pH 2.65}$$
$$k = 43.2 \times 10^{-5}\ s^{-1} \qquad \text{at pH 2.05}$$

INFRARED SPECTROMETRY

The infrared method is less exact. The thickness of the layer is less reproducible. In most cases the solvents absorb, and the concentrations needed are larger. In some cases, however, infrared spectrophotometry is particularly useful, such as when an absorption band is available in an "empty" region of the spectrum, e.g., the Sn—H stretching frequency of a trialkyltin hydride, R_3SnH, at 1800 cm^{-1} and the N═C stretching frequency of an isocyanide, RN═C, at 2150 cm^{-1}.

The logarithmic form of the first-order rate equation is

$$\log c_t = \log c_{t=0} - 0.434kt$$

If $\log c_t$ is plotted versus time, the points are situated on a straight line with a slope of -0.434k. Thus, to calculate the rate constant, it is not necessary to know the initial concentration.

Where an analytical function ΔF is known to be proportional to the concentration c, then a straight line is also formed with slope -0.434k when $\log \Delta F$ is plotted versus time, in accordance with

$$\log \Delta F_t = \log c_t + \text{constant}$$

Consequently, for first-order reactions it is not necessary to measure directly the actual concentration of one of the reactants as a function of time. It is sufficient to pursue the change in a physical parameter ΔF, the decrease of which is linearly proportional to the progress of the reaction process. When the reactant concentrations are sufficiently low, of the order of 0.1 mol/liter or lower, many physical parameters obey this condition. See also p. 00 in connection with Guggenheim's method.

For example, the change in absorbance in the ultraviolet or infrared at a special wavelength or in the refractive index of the solution may serve as this physical parameter.

DILATOMETRY

Another parameter is the volume change of a reacting liquid, which can be followed using a dilatometer. The requirement of constant temperature in dilatometric procedures is higher than for most kinetic runs, e.g., ±0.003 K. Otherwise, the dilatometer functions as a thermometer. Correction for temperature fluctuations is possible when two dilatometers of the same size are used; one dilatometer is filled with the reacting solution and the other with pure solvent. See Fig. 3.2. When the surface height in the capillary, h, decreases as a function of time, the difference $h_t - h_{t=\infty}$ is proportional to the conversion and

$$\ln(h_t - h_{t=\infty}) = \text{constant} - kt$$

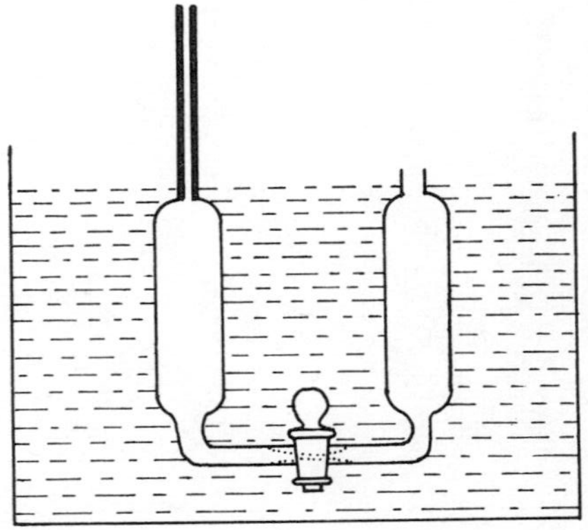

FIG. 3.2 Constant-temperature bath.

The rate constant is calculated from a plot of ln or of $\log(h_t - h_{t=\infty})$ versus time.

BARIMETRY (MEASUREMENT OF PRESSURE CHANGE)

The change in pressure can be measured when volatile products are formed in the course of reaction. An example is the thermolysis reaction of a 1-alkynyl ether during which isobutylene is formed;

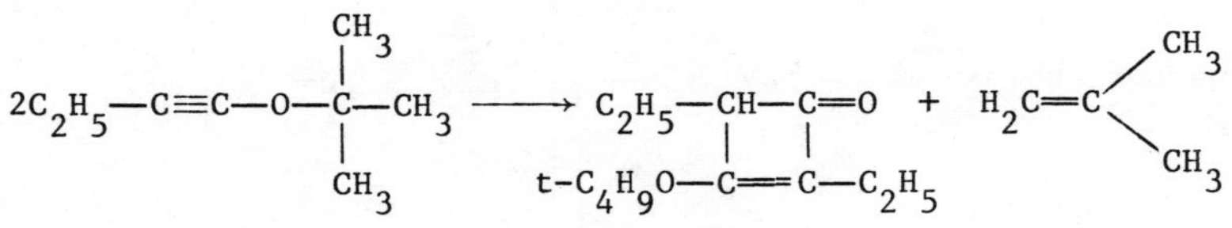

The reaction was performed in a closed system at a temperature of 84.5°C; the initial concentration was 0.12 mol/liter in decalin as solvent. The pressure p was measured as a function of time. At the end of the reaction the pressure held constant at p_∞:

$$\log(p_\infty - p) = \text{constant} - 0.434(2k)t$$

The coefficient 2 in the last term arises from the fact that for the formation of each molecule of isobutylene two molecules of ether are needed.

$$k = 8.73 \times 10^{-5}\ s^{-1}$$

Instead of measuring the change in pressure at constant volume, the change in volume at constant pressure may also be pursued.

The latter method is not as suitable, however, because isobutylene is distributed between the gas and liquid phases. According to Henry's law, the ratio of the concentrations in the gas and liquid phases is constant.

$$\frac{p}{c} = \text{constant}$$

When the isobutylene pressure is constant, the concentration of isobutylene in the liquid phase, c_L, is also a constant. The total amount of isobutylene is $c_L V_L + c_g V_g$, and therefore it is not proportional to the measured gas volume, V_g. On the other hand, if the volume is constant, then the concentrations of isobutylene in liquid and gas phases are proportional. Now, the total amount of isobutylene is proportional to the isobutylene pressure.

$$\underset{\text{I}}{C_2H_5C{\equiv}C{-}OC(CH_3)_3} \xrightarrow{k_1} \underset{\text{II}}{C_2H_5\overset{H}{C}{=}C{=}O} + (CH_3)_2C{=}CH_2$$

$$\underset{\text{II}}{C_2H_5\overset{H}{C}{=}C{=}O} + \underset{\text{I}}{C_2H_5C{\equiv}C{-}OC(CH_3)_3} \xrightarrow{k_2} \text{cyclobutenone}$$

The reverse reaction of the first step is neglected because the isobutylene is predominantly in the gas phase. The reverse reaction of the second step can also be neglected since the cyclobutenone does not decompose under the reaction conditions. The ketone is a very reactive intermediate which attains an extremely low, almost constant concentration in the solution. Thus, the stationary-state method can be applied:

$$\frac{dc_{II}}{dt} = k_1 c_I - k_2 c_{II} c_I = 0$$

$$c_{II} = \frac{k_1 c_I}{k_2 c_I} = \frac{k_1}{k_2}$$

The rate of disappearance of the initial compound is

$$\frac{-dc_I}{dt} = k_1 c_I + k_2 c_{II} c_I = k_1 c_I + k_1 c_I = 2k_1 c_I$$

Integration:

$$c_I = c_{I,t=0} \exp(-2k_1 t)$$

The analogous reaction in the presence of aniline is

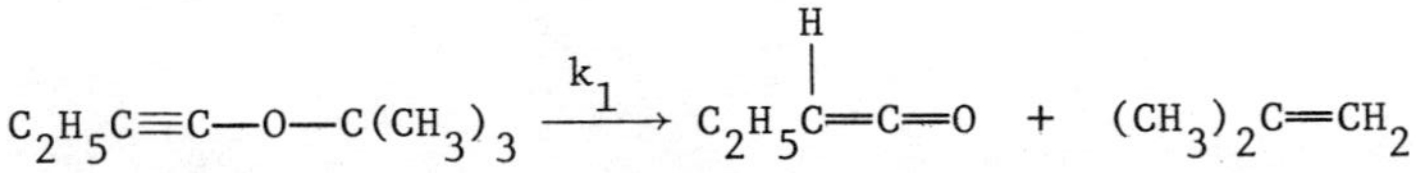

$$C_2H_5C{\equiv}C{-}O{-}C(CH_3)_3 \xrightarrow{k_1} C_2H_5\overset{\overset{\textstyle H}{|}}{C}{=}C{=}O + (CH_3)_2C{=}CH_2$$

$$\text{I} \qquad\qquad\qquad \text{II}$$

$$C_2H_5\overset{\overset{\textstyle H}{|}}{C}{=}C{=}O + C_6H_5NH_2 \xrightarrow{k_2'} C_2H_5\overset{\overset{\textstyle H_2}{}}{C}{-}\overset{\overset{\textstyle O}{\|}}{C}{-}NHC_6H_5$$

$$\text{II}$$

Ketene is much more reactive toward aniline than butynyl ether; i.e., $k_2' \gg k_2$. The rate of disappearance of the initial compound in this case is now

$$v = k_1 c_I$$

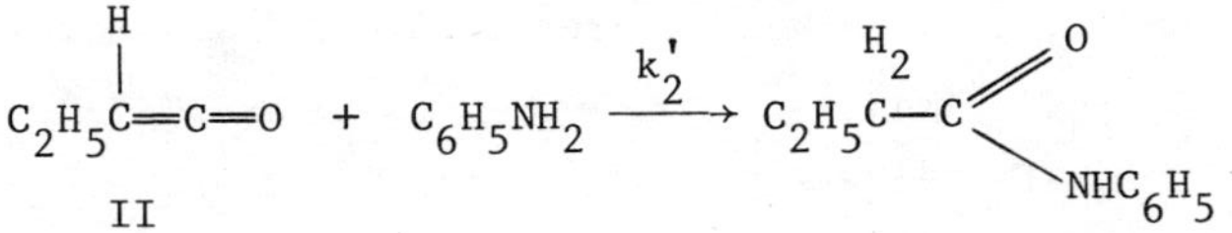

and the rate equation becomes

$$c_I = c_{I,t=0} \exp(-k_1 t)$$

Moreover, the amount of isobutylene formed per mole of ether I reacted is now twice as great as previously. It should be obvious that the determination of k_1 in the latter reaction permits a check on the value of k_1 determined in the absence of aniline.

Manometric determinations of the rate of oxygen consumption by bacteria and other biological materials are often performed in a Warburg apparatus. Such an apparatus can also be useful in nonbiological experiments.

POTENTIOMETRY

Potentiometric determinations of rates are sometimes possible in, e.g., bromination reactions:

$$Br_2 + 2e \rightleftarrows 2Br^-$$

The redox potential can be measured with a platinum electrode with a reference calomel electrode. According to Nernst, the potential of the system is equal to

$$E = \text{constant} + \frac{RT}{2F} \ln \frac{c_{Br_2}}{c_{Br}^2} = \text{constant}' + 0.0295 \log c_{Br_2}$$

when an almost constant excess of bromide ions is present and when the reaction is carried out at 25°C. When bromine is consumed in a first-order or even a pseudo-first-order reaction

$$\log c_{Br_2,t} = \log c_{Br_2,t=0} - 0.434kt$$

and thus

$$E = \text{constant}'' - 0.0295 \times 0.434kt$$

When plotting the potential E versus time t, e.g., on a recorder, a straight line results. The rate constant can be calculated from the slope of this line. This method has been applied in the bromination of N,N-dialkylanilines.

R. P. Bell and P. De Maria, J. Chem. Soc., B 1958:161; R. P. Bell and E. N. Ramsden, J. Chem Soc., B 1969:1057.

J. Dubois, R. Uzan, and P. Alcais, Bull. Soc. Chim. Fr. 1968:617.

CONDUCTOMETRY

When ions are produced or consumed the rate can be obtained from conductivity data. For high precision it is not always permissible to assume a linear correlation between conductivity and concentration. Moreover, by minimizing errors inherent in the values of solvent composition, temperature, electrical resistivity, and time, Shiner et al. were able to determine solvolysis rate constants with very high precision. This precision was required for the determination of secondary deuterium isotope effects (see pages 128 and 132).

B. L. Murr and V. J. Shiner, J. Amer. Chem. Soc. 84:4672 (1962).

See also R. N. McDonald and G. E. Davis, J. Org. Chem. 38:138 (1973).

POLARIMETRY

For chiral compounds an obvious physical parameter is optical rotation. Classical examples are the stereochemical investigations of S_N2 processes, e.g., the reaction between iodide ions and optically active 2-iodooctane in acetone as solvent. When sodium iodide is added to a

$$I^- + C_6H_{13}-\overset{\overset{\displaystyle H}{|}}{\underset{\underset{\displaystyle CH_3}{|}}{C}}-I \;\rightleftharpoons\; I-\overset{\overset{\displaystyle H}{|}}{\underset{\underset{\displaystyle CH_3}{|}}{C}}-C_6H_{13} + I^-$$

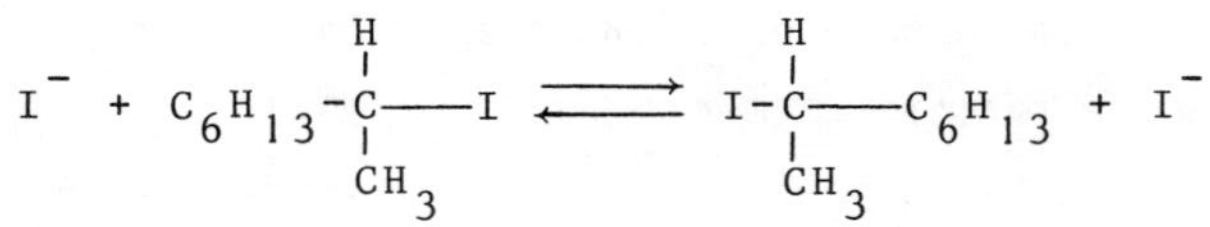

solution of optically active 2-iodooctane, the optical activity declines with the passage of time. The rate of decrease of rotation, i.e., the rate of racemization, is twice the rate of inversion, because one inversion process $\underline{R} \to \underline{S}$ increases the amount of racemic material by two molecules, an $\underline{R}$ and an $\underline{S}$. Or, stated otherwise, every molecule of laevorotatory $\underline{R}$ converted to dextrorotatory $\underline{S}$ effectively cancels the contributions of two molecules of $\underline{R}$ to the total optical activity of the solution in the polarimeter. Thus, the rate of decline of optical activity, i.e., the rate of racemization, takes place with a velocity which is twice as great as the conversion of $\underline{R}$ to $\underline{S}$. The rate of substitution was obtained from experiments with radioactively labeled iodide ions. The NaI had been irradiated for some time with neutrons:

$$I^{*-} + RI \underset{k}{\overset{k}{\rightleftharpoons}} I^- + RI^*$$

E. D. Hughes, F. Juliusburger, S. Masterman, B. Tonley, and J. Weiss, J. Chem. Soc. 1935:1525.

$\underline{R}$

$\underline{S}$

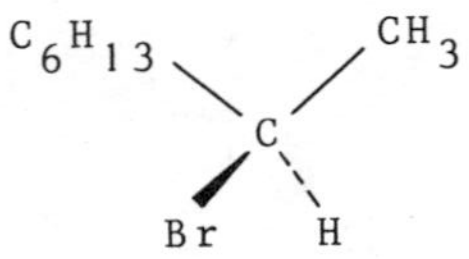

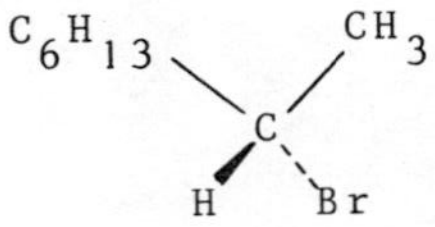

Initially, all activity resides in the iodide ions. If the active fraction is called β and the total iodide concentration B, $[I^{*-}] = B\beta$ and $[I^-] = B(1 - \beta)$. After a while the activity is distributed between iodide ions and organic iodide. Let the concentration of radioactive organic iodide at time t be x. Furthermore, let the total concentration of organic iodide ($[RI] + [RI^*]$) be A. Then, at time t, $[I^{*-}] = B\beta - x$, $[I^-] = B(1 - \beta) + x$, $[RI^*] = x$, and $[RI] = A - x$.

$[RI^*]$ increases by reaction of I^{*-} with RI and decreases by reaction of I^- with RI^*. Thus,

$$\frac{dx}{dt} = k(B\beta - x)(A - x) - k[B(1 - \beta) + x]x$$
$$= k[B\beta A - (A + B)x]$$

Integration:

$$\frac{-1}{A + B} \ln[-(A + B)x + B\beta A] = kt + \text{constant} \qquad \text{for } t = 0,\ x = 0$$

and

$$\text{Constant} = \frac{-1}{A + B} \ln(B\beta A)$$

Thus,

$$\frac{-1}{A + B} \ln \frac{-(A + B)x + B\beta A}{B\beta A} = kt$$

x is measured by the ratio of the activity of the organic iodide to that of the inorganic iodide:

$$\gamma = \frac{[RI^*]/A}{[I^{*-}]/B} = \frac{B}{A}\,\frac{x}{B\beta - x}$$

Rates of substitution and inversion appeared to be equal within experimental error. Thus, every act of substitution was accompanied by inversion of configuration at the carbon center of reaction.

In a similar investigation of an S_N2 reaction at a primary carbon atom monodeuterated butyl bromide was used, which has a small, but measurable optical activity. Also in this S_N2 process, the reaction of butyl bromide with bromide ions, the rate of inversion was found to be equal to the rate of substitution. The latter had been measured with radioactive bromide ions.

A. Streitwieser, J. Amer. Chem. Soc. 75:5014 (1953).

C_3H_7CHDBr

L. J. Le Roux and S. Sugden, J. Chem. Soc. 1939:1279.

MASS SPECTROMETRY

In reactions involving isotopes, the mass spectrometer can be used for analytical pursuit--for example, the decomposition of tert-butyldimethylsulfonium iodide in water at 59°C:

$$(CH_3)_3C\overset{+}{S}(CH_3)_2 + I^- \longrightarrow (CH_3)_3CI + S(CH_3)_2$$

$$(CH_3)_3COH \quad \text{and} \quad [(CH_3)_2C{=}CH_2 + HI] \longleftarrow (CH_3)_3CI$$

W. H. Saunders and S. Ašperger, J. Amer. Chem. Soc. 79:1612 (1957).

The slow step in this reaction is the ionization of a C—S bond:

$$\text{Tert-butyl-}\overset{+}{S}(CH_3)_2 \longrightarrow \text{tert-butyl}^+ + S(CH_3)_2$$

Sulfur consists of 95% ^{32}S, 4% ^{34}S, and 1% ^{33}S. The latter isotope can be neglected. In the slow step two different reactions are possible, the breaking of the $^{32}S—C$ and of the $^{34}S—C$ covalencies. For reasons to be discussed in a subsequent section of this book (Chap. 5) the breaking of the bond with the heavier isotope is somewhat more difficult. Thus, the $^{32}S—C$ is more readily ruptured than the $^{34}S—C$ covalency, and we would anticipate that $k_{^{32}S} > k_{^{34}S}$ if this bond breaking is occurring in the rate-determining step. In the initial few percent of the reaction, the ^{32}S in the product dimethyl sulfide is increased beyond its normal abundance of ca. 95% in the reactant sulfonium ion, as measured by mass spectrometry:

$$\frac{k_{^{32}S}}{k_{^{34}S}} = \frac{(^{32}S/^{34}S)_{products}}{(^{32}S/^{34}S)_{reactants}} = 1.018$$

As will be shown in Chap. 5, if the rate-determining step had involved increased bonding to sulfur, the isotope effect would have been measurably smaller.

NUCLEAR MAGNETIC RESONANCE

Quantitative measurements can also be performed with a proton magnetic resonance spectrometer. For example, the proton magnetic resonance spectrum of tert-butylacetylene shows two signals, a large signal at higher field and a small signal at lower field corresponding to the nine tert-butyl hydrogens and the single acetylenic hydrogen, respectively. The intensity ratio is 9:1. When the substrate is dissolved in a mixture of acetone, D_2O, and a little KOD, the acetylenic proton signal gradually decreases in intensity:

$$(CH_3)_3CC{\equiv}CH + D_2O \overset{OD^-}{\rightleftarrows} (CH_3)_3CC{\equiv}CD + HDO$$

Because of the large excess of D_2O, this equilibrium lies far to the right. The rate of this exchange can be determined by measuring the height of the signal as a function of time. In acetone + 10 vol % D_2O at approximately 25°C,

$$v = kc_{acetylene}c_{OD^-} \qquad \text{with } k = 2.1 \text{ liters mol}^{-1}\ s^{-1}$$

This k is the rate constant of the reaction

$$(CH_3)_3CC{\equiv}CH + OD^- \longrightarrow (CH_3)_3CC{\equiv}C^- + HDO$$

TREATMENT OF DATA: GUGGENHEIM'S METHOD

The reaction rate constant of a first-order or a pseudo-first-order reaction can be calculated by measuring the decrease or increase of a physical parameter as a function of time (page 51). In general the so-called infinity value of the physical parameter will be necessary. The infinity value usually means the value after 8 or 10 half-lives. Sometimes consecutive reactions or an insufficiently stable instrument make it impossible to measure a reliable infinity value. In such a case the rate constant can be obtained by the familiar Guggenheim method if during a period of 3 or 4 half-lives reliable measurements are possible.

E. A. Guggenheim, Phil. Mag. 2:538 (1926).

Let us take as an example the dilatometric method (page 52). For the first-order or pseudo-first-order reaction process the kinetic equation is

$$\ln(h_t - h_{t=\infty}) = \text{constant} - kt \tag{3.1}$$

$$h_t - h_{t=\infty} = \text{constant}' \exp(-kt) \tag{3.2}$$

where h_t as usual represents the observed dilatometer height at time t.

Dilatometric observations are made at regular intervals so that the series of measurements are separable into two groups corresponding to times

$$t_1, t_2, t_3, \ldots$$

and

$$t_1 + \Delta t, t_2 + \Delta t, t_3 + \Delta t, \ldots$$

Moreover, the magnitude of Δt must be at least two half-lives for maximum precision of the results:

$$h_{t_i} - h_{t=\infty} = \text{constant}' \exp(-kt_i)$$

$$h_{t_i+\Delta t} - h_{t=\infty} = \text{constant}'[\exp -k(t_i + \Delta t)]$$

$$h_{t_i} - h_{t_i+\Delta t} = \text{constant}'[1 - \exp(-k\,\Delta t)] \exp(-kt_i) \tag{3.3}$$

Since the term $1 - \exp(-k\,\Delta t)$ is constant during a run, a plot of

$$\ln(h_{t_i} - h_{t_i+\Delta t})$$

versus time gives a straight line with slope $-k$. Clearly, the rate constant calculated from this exercise has no reference to either the initial or infinity values of concentration of the reactants or products.

COMPUTERIZED DATA REDUCTION

A computer-based alternative to the Guggenheim treatment is often used when reliable zero and infinity values are not at hand and when, in addition, least-squares processing of the results is desired to attain the most precise value of the rate constant. The LSKINI program (De Tar), which is widely applied for these purposes, stems from the following considerations.

D. F. DeTar, Computer Programs for Chemistry, Vol. 1 (Benjamin, New York, 1968), p. 126, and Vol. 4 (Academic Press, New York, 1972), p. 1.

According to Eq. (3.1),

$$h_{t_1} - h_{t_1+\Delta t} = \text{constant}'[1 - \exp(-k\ \Delta t)]\ \exp(-kt_1) \qquad (3.4)$$

and

$$h_{t_2} - h_{t_2+\Delta t} = \text{constant}'[1 - \exp(-k\ \Delta t)]\ \exp(-kt_2) \qquad (3.5)$$

Dividing (3.4) by (3.5) results in

$$\frac{h_{t_1} - h_{t_1+\Delta t}}{h_{t_2} - h_{t_2+\Delta t}} = \exp[k(t_2 - t_1)] \qquad (3.6)$$

which is transformed to

$$k = \frac{1}{t_2 - t_1} \ln \frac{h_{t_2} - h_{t_1+\Delta t}}{h_{t_2} - h_{t_2+\Delta t}} \qquad (3.7)$$

There are similar equations for t_3, etc.

The LSKINI program solves Eq. (3.7) for k using a least-squares treatment in which either a scalar or relative error in h may be minimized; it has, in fact, capabilities for making least-squares adjustments to any or all the parameters of Eq. (3.7).

4 Theory of Reaction Rates

How to analyze rate constant data

As is well known a reaction rate generally increases with temperature, often some 10 to 15% per degree centigrade. What is the origin of this generality? It could be the increase in kinetic energy of the molecules. The average kinetic energy of translation is (3/2)kT. Its increase at 300 K is (1/3)% per degree centigrade, much less than the 10 or 15% observed for the reaction rate. Clearly there must be another reason for this increase than simply the kinetic energy.

K = degree Kelvin

ARRHENIUS EQUATION

In 1889, Arrhenius assumed that a reaction $A \rightarrow B$ takes place by way of an activated intermediate A^* that is in equilibrium with A. For this equilibrium

$$\frac{[A^*]}{[A]} = K = \frac{k_1}{k_2} \qquad (4.1)$$

$$A \underset{k_2}{\overset{k_1}{\rightleftarrows}} A^* \xrightarrow{k_3} B$$

Arrhenius had to assume that the concentration of A* was much smaller than the concentration of A because an intermediate A* had never been isolated. The rate of formation of product B is therefore

$[A^*] << [A]$

$$k_3[A^*] = k_3K[A] \qquad (4.2)$$

The empirical rate equation is $v = k[A]$, so that $k = k_3K$. Van't Hoff's equation relating the equilibrium constant to temperature, sometimes called the Van't Hoff isochore, is

$$\frac{d \ln K}{dT} = \frac{E}{RT^2} \qquad (4.3)$$

Consequently,

$$\frac{d \ln(k/k_3)}{dT} = \frac{E}{RT^2} \quad \text{or} \quad \frac{d \ln k}{dT} - \frac{d \ln k_3}{dT} = \frac{E}{RT^2} \qquad (4.4)$$

Arrhenius suggested that k_3 is independent of the temperature so that $(d \ln k_3)/dT = 0$. Thus

$$\frac{d \ln k}{dT} = \frac{E}{RT^2} \qquad (4.5)$$

which on integration gives

$$k = A \exp \frac{-E}{RT} \qquad (4.6)$$

This equation has considerable empirical justification, since a linear relation is most commonly observed between ln k and 1/T. However, it is not clear why in every reaction an intermediate A* occurs and why k_3 is independent of the temperature, whereas k_1 and k_2 are temperature dependent.

Instead of using the concentration of a highly energetic intermediate A*, it is much more satisfactory to use the number of particles with velocities exceeding a given value (see Fig. 4.1). This number rapidly increases with temperature. The energy distribution is related to the collisions between molecules.

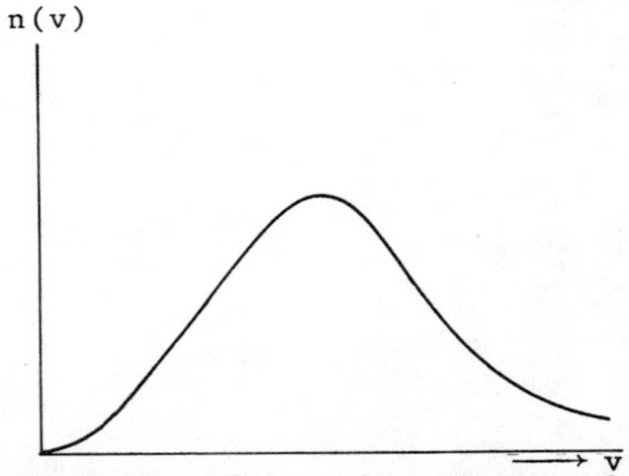

FIG. 4.1 Velocity distribution after Maxwell.

This idea was first put forward in 1922 by Lindemann, who made the following suggestion: In general, the energy required for a molecule to undergo a unimolecular thermal reaction is not to a sufficient extent present in the molecule itself but comes from collisions. Collision involves a second particle. Therefore, the rate of reaction should be second order in A. In gas phase reactions at very low pressures this behavior has indeed been observed. But except at these low pressures the observed kinetics are first order, because under normal conditions frequent collisions occur and the energized molecules A* are in rapid equilibrium with A. Theoretical expressions for the rates of unimolecular gas phase reactions were developed by Lindemann and Hinshelwood. A complicating factor is that the energy acquired by the molecule may be distributed among a considerable number of degrees of freedom. The theory was extended by Rice, Ramsperger, and Kassel. The expression RRK

$$A \longrightarrow \text{products}$$

$$A + A \longrightarrow A^* + A$$

$$A \rightleftharpoons A^* \xrightarrow{\text{slow}} \text{products}$$

For a more elaborate treatment, see, e.g., R. P. Wayne, in Vol. 2, The theory of kinetics, Comprehensive Chemical Kinetics, C. H. Bamford and C. F. H. Tipper, eds. (Elsevier, Amsterdam, 1969), Chap. 3.

theory is derived from the initials of these authors. An alternative theory was developed by Slater.

These collision theories on gas phase reactions are not easy to apply to--generally complicated--molecules of organic compounds. Moreover, most reactions in organic chemistry take place in the liquid phase, and therefore we shall make only sparing use of collision theory.

TRANSITION-STATE THEORY

Transition-state theory has been found, on the other hand, to be much more widely applicable to the analysis of organic reactions in solution. In this concept a reaction step is viewed at three different stages along its path, an initial state, a transition state, and a final state. Reaction path means the route of minimum Gibbs function. Such a route is visualized by a contour line figure (see Fig. 4.2). The corresponding bimolecular reaction can be expressed as

$$XY + Z \longrightarrow X + YZ$$

In the initial state, the system resides in valley a and in the final state in valley b. The system moves from a to b via the equivalent of a mountain pass connecting two valleys. The dashed-line path is called the reaction coordinate. Figure 4.3 traces the behavior of the Gibbs function along the reaction coordinate.

The presence of such a mountain can be elucidated as follows: The Morse curve, which gives the relationship between the potential energy of the molecule XY as a function of the distance r_{XY} between X and Y,

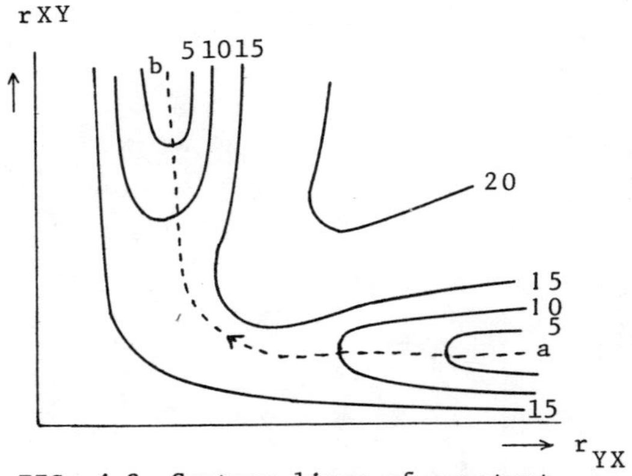

FIG. 4.2 Contour lines of constant Gibbs function. The numbers represent values of G for each contour line. r_{XY} = the distance between X and Y; r_{YZ} = the distance between Y and Z.

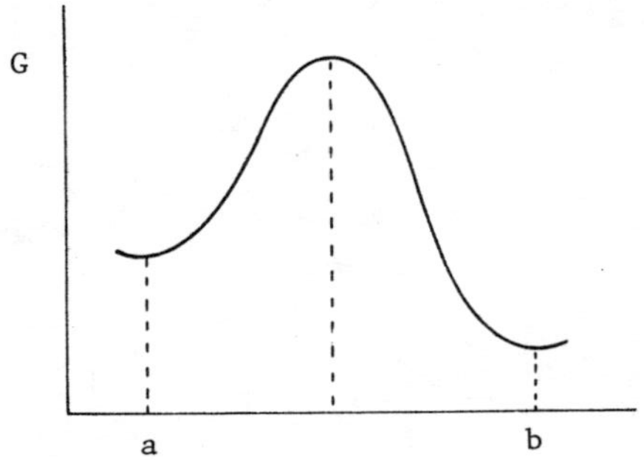

FIG. 4.3

is shown in Fig. 4.4. Figure 4.5 shows the combination of the XY-curve with that of the molecule YZ, the Y-Z distance r_{YZ} increasing to the left. During the reaction, the course of the potential energy variation follows the combination of both curves. The region around the intersection maximum corresponds to the location of the transition state. A molecule or combination of molecules in this state is called the activated complex. It is to be emphasized, however, that the transition state is at an energy minimum with respect to the direction perpendicular to the reaction path.

Monomolecular reactions also can be considered to take place by way of an activated complex:

$$XY \longrightarrow XY^{\neq} \longrightarrow X + Y$$

Generally, energy will be necessary before XY can react. $XY^{\neq}$ is the activated complex, i.e., the species at the reaction path position where the potential energy reaches its maximum value. Transition-state theory is based on the assumption of an equilibrium between the molecular structure of the initial state and the molecule-like configuration of the activated state. Emphasis must be given to the fact that the latter is a state being treated as if it were that of a molecule in which a critical bond is in the process of breaking. The significant difference is that a normal molecular state of XY lies in a potential energy well, whereas $XY^{\neq}$ is located at the top of a barrier, i.e., a potential energy maximum with respect to the direction of the reaction path.

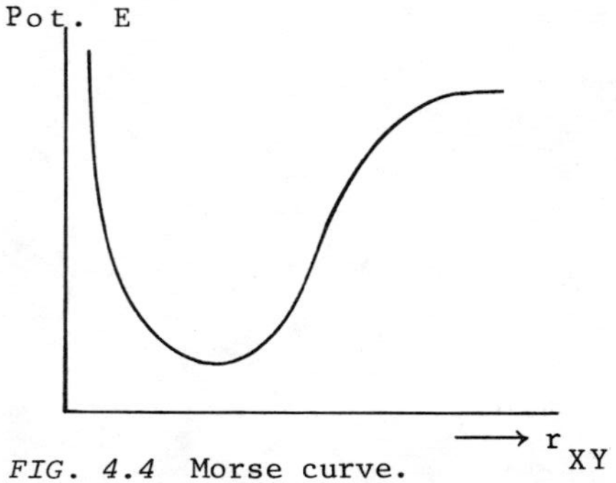

FIG. 4.4 Morse curve.

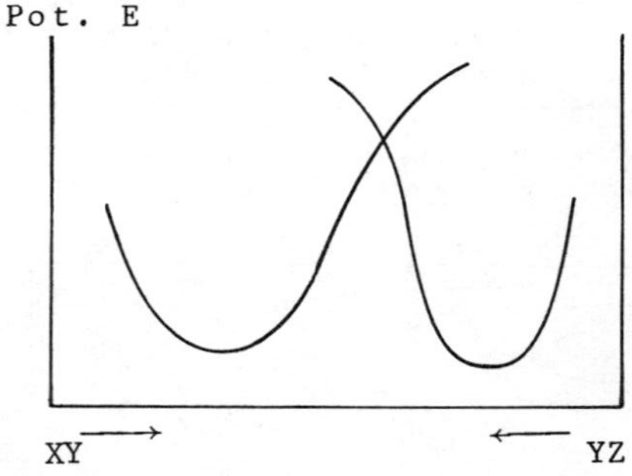

FIG. 4.5 Merger of Morse curves.

$$XY \rightleftharpoons XY^{\neq}$$

The equilibrium constant

$$K = \frac{[XY^{\neq}]}{[XY]} \tag{4.7}$$

How does the equilibrium constant K depend on the molecular parameters? To answer this question we must resort to statistical thermodynamics. Consider the equilibrium $A \rightleftharpoons B$. Given that the energy of molecule A is ε_A and that of molecule B is ε_B, it follows from the Boltzmann equation that the distribution of molecules between the states of A and B is equal to

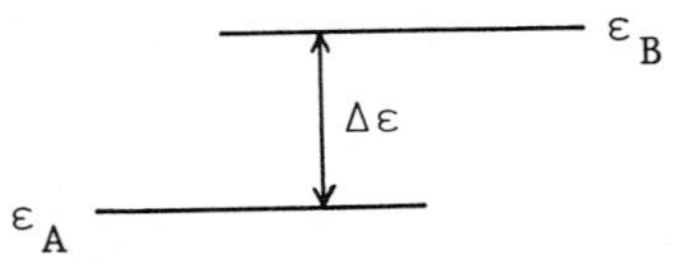

$$\frac{\exp(-\varepsilon_B/kT)}{\exp(-\varepsilon_A/kT)} = \exp \frac{-(\varepsilon_B - \varepsilon_A)}{kT}$$

$$= \exp \frac{-\Delta\varepsilon}{kT} \tag{4.8}$$

and

$$K = \frac{[B]}{[A]} = \exp \frac{-\Delta\varepsilon}{kT} \tag{4.9}$$

This treatment is somewhat oversimplified because it does not take into account the fact that molecules A and B can appear in energetic states which differ as to translational, vibrational, and rotational degrees of freedom:

For A: $\varepsilon_A^{(0)} \varepsilon_A^{(1)} \varepsilon_A^{(2)} \ldots$

For B: $\varepsilon_B^{(0)} \varepsilon_B^{(1)} \varepsilon_B^{(2)} \ldots$

The probability of occupying these states is proportional to $\exp\left(-\varepsilon_A^{(0)}/kT\right)$, $\exp\left(-\varepsilon_A^{(1)}/kT\right)$, $\exp\left(-\varepsilon_A^{(1)}/kT\right)$, etc. Thus,

$$K = \frac{\sum_i \exp\left(-\varepsilon_B^{(i)}/kT\right)}{\sum_j \exp\left(-\varepsilon_A^{(j)}/kT\right)}$$

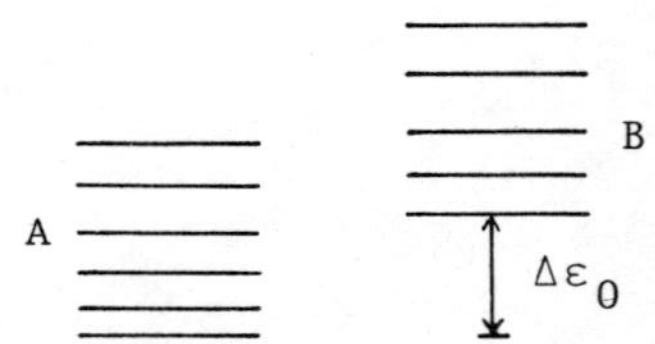

Energy levels

where $\sum_i \exp\left(-\varepsilon^{(i)}/kT\right)$ is called the partition function, abbreviated as Q, so that

$$K = \frac{Q_B}{Q_A} \qquad (4.10)$$

In this derivation A and B have a common energy origin. However, it is customary to consider the lowest energy level of each molecule as its zero-point energy level, and therefore

$$K = \frac{Q_B}{Q_A} \exp \frac{-\Delta\varepsilon_0}{kT} \qquad (4.11)$$

SIMPLE UNIMOLECULAR TREATMENT

For the monomolecular reaction

$$XY \overset{K}{\rightleftarrows} XY^{\neq} \rightleftarrows \text{products}$$

we have $K = (Q_{XY^{\neq}}/Q_{XY}) \exp(-\Delta\varepsilon_0/kT)$, and for a bimolecular reaction, derived analogously,

$$XY + Z \underset{\longleftarrow}{\overset{K}{\longrightarrow}} XYZ^{\neq} \underset{\longleftarrow}{\longrightarrow} X + YZ$$

we have

$$K = \frac{Q_{XYZ^{\neq}}}{Q_{XY}Q_Z} \exp \frac{-\Delta\varepsilon_0}{kt} \qquad (4.12)$$

where $\Delta\varepsilon_0 = \varepsilon_0^{(\neq)} - \left(\varepsilon_0^{(XY)} + \varepsilon_0^{(Z)}\right)$.

To a good approximation the translational, vibrational, and rotational energies can be regarded as being independent of each other:

$$\varepsilon_{total} = \varepsilon_{transl} + \varepsilon_{vibr} + \varepsilon_{rot}$$

and

$$\exp \frac{-\varepsilon}{kT} = \exp \frac{-\varepsilon_{transl}}{kT} \exp \frac{-\varepsilon_{vibr}}{kT} \exp \frac{-\varepsilon_{rot}}{kT} \qquad (4.13)$$

and

$$Q = Q_{transl}Q_{vibr}Q_{rot}$$

A nonlinear n-atomic molecular array has three degrees of freedom for translation along the coordinates x, y, and z, three for rotation around these coordinate axes, and 3n - 6 for vibration,

$$Q_{transl} = Q_t^x Q_t^y Q_t^z \qquad Q_{rot} = Q_r^x Q_r^y Q_r^z \qquad Q_{vibr} = Q_v^{(1)} Q_v^{(2)} Q_v^{(3)} \ldots$$

since for each degree of freedom we require a Q.

A linear molecule has two degrees of freedom for rotation and $3n - 5$ for vibration. The masses of XY and $XY^{\neq}$ are equal, so their Q_{transl} are also equal. Q_{rot} depends on the moments of inertia of the molecules. When the molecule XY consists of several atoms, the moments of inertia change relatively little when the bond X—Y is stretched. Therefore, we only need to consider Q_{vibr}. This Q_{vibr} is the product of the partition functions of the various normal vibrational modes. Let us assume that in the conversion of XY to $XY^{\neq}$ the only important change is in the stretching of the X—Y bond. Then we can restrict the discussion to the vibrational mode in which X—Y is undergoing this stretching motion. The partition function of a vibration is defined as

$$\sum_i \exp \frac{-\varepsilon_{vibr}^{(i)}}{kT} \qquad (4.14)$$

the sum being over all vibrational levels. When the vibration is harmonic the energy levels are given by

$$\varepsilon_n = \left(n + \frac{1}{2}\right) h\nu, \qquad n = 0, 1, 2, \ldots \qquad (4.15)$$

ν is the frequency of the vibration and h is Planck's constant. From (4.14) and (4.15)

$$Q_{vibr} = \sum_n \exp \frac{-[n + (1/2)]h\nu}{kT} \qquad (4.16)$$

The sum of this infinite geometric progression is equal to

$$Q_{vibr} = \exp \frac{-(1/2)h\nu}{kT}\left(1 - \exp \frac{-h\nu}{kT}\right)^{-1} \tag{4.17}$$

Usually, for vibrations $h\nu >> kT$; thus

$$Q_{XY} \simeq \exp \frac{-(1/2)(h\nu)}{kT} \tag{4.18}$$

In the transition state the stretching vibration of X—Y is in the art of becoming a translation. Still regarding it as a sort of (imaginary) vibration, we can, under these circumstances, assume that $h\nu << kT$. In this event $\exp[-(1/2)h\nu^{\neq}/kT] \simeq 1$ and $[1 - \exp(-h\nu^{\neq}/kT)]^{-1} \simeq kT/h\nu^{\neq}$, so that the pertinent partition function of the transition state

$$Q_{XY^{\neq}} = \frac{kT}{h\nu^{\neq}} \tag{4.19}$$

For the equilibrium

$$XY \underset{}{\overset{K}{\rightleftarrows}} XY^{\neq}$$

we have

$$K = \frac{kT}{h\nu^{\neq}} \exp \frac{(1/2)h\nu}{kT} \exp \frac{-\Delta\varepsilon_0}{kT} = \frac{kT}{h\nu^{\neq}} \exp \frac{-[\Delta\varepsilon_0 - (1/2)h\nu]}{kT} \tag{4.20}$$

The velocity v with which $XY^{\neq}$ proceeds to full decomposition is given by the product of $[XY^{\neq}]$ and the probability that the complex $XY^{\neq}$

$\exp(-a) \simeq 1$ for $a \to 0$

Note: This assumption will not always be applicable.

$$XY^{\neq} \longrightarrow \text{products}$$

will decompose. This probability can be characterized by the frequency $\nu^{\neq}$, because almost as soon as X and Y move apart, their separation results in complete reaction. Thus,

$$v = \nu^{\neq}[XY^{\neq}] = \nu^{\neq}K[XY] = \frac{kT}{h} \exp \frac{-\Delta\varepsilon_0'}{kT} [XY]$$

where $\Delta\varepsilon_0'$ is equal to $\Delta\varepsilon_0$ corrected for the zero-level energy of XY. For the rate constant k_r of a first-order reaction in the empirical equation

$$v = k_r[XY]$$

it follows that

$$k_r = \frac{kT}{h} \exp \frac{-\Delta\varepsilon_0'}{kT} \qquad (4.21)$$

MORE GENERAL TREATMENT

The following treatment is more general and readily applied to both unimolecular and bimolecular rate processes; we shall focus on the critical bond undergoing conversion to a translation in the activated complex:

$$XY \longrightarrow X + Y$$
$$XY + Z \longrightarrow X + YZ$$

The derivation given by H. Eyring et al. for a bimolecular process also starts from an equilibrium:

W. S. K. Wynne-Jones and H. Eyring, J. Chem. Phys. 3:107, 492 (1935).

$$\text{initial reactants} \rightleftharpoons \text{activated complex} \rightleftharpoons \text{products}$$

In the present case

$$XY + Z \rightleftharpoons XYZ^{\neq} \rightleftharpoons X + YZ$$

For the equilibrium between reactants and activated complex the equilibrium constant is K with the proviso that when activity constants have to be taken into account Eq. (4.36) must be used; see page 93.

$$K = \frac{[XYZ^{\neq}]}{[XY][Z]}$$

The reaction consists mainly of the transfer of Y from X to Z. In the course of reaction the vibration between X and Y becomes a translation. What is the average translational velocity of Y in this motion? According to the Boltzmann equation, the probability for a particle of mass m finding itself in a state with a translational velocity v is proportional to $\exp(-mv^2/2kT)$. The average velocity in one direction $\bar{v}$, for example, to the right, is given by

Translational energy = $(1/2)mv^2$

$$\bar{v} = \frac{\int_0^{\infty} \exp(-mv^2/2kT)v\ dv}{\int_0^{\infty} \exp(-mv^2/2kT)\ dv}$$

which is equal to

$$\bar{v} = \left(\frac{2kT}{\pi m}\right)^{1/2} \qquad (4.22)$$

Proof: The numerator can be written as

$$\left(\frac{2kT}{m}\right) \int_0^{\infty} \exp(-x^2)x\ dx$$

and the denominator as

$$\left(\frac{2kT}{m}\right)^{1/2} \int_0^{\infty} \exp(-x^2)\ dx$$

where $x = (m/2kT)^{1/2}v$. Since $\int_0^{\infty} \exp(-x^2)x\ dx = 1/2$ and $\int_0^{\infty} \exp(-x^2)\ dx = \sqrt{\pi}/2$,

$$\bar{v} = \left(\frac{2kT}{m\pi}\right)^{1/2}$$

The assumption stated previously is that the system is in equilibrium and that the particles move via the transition state from the left to the right and from the right to the left. The concentration of particles in the transition state is $[XYZ^{\neq}]$. Half of the particles, $(1/2)[XYZ^{\neq}]$, move to the right with an average velocity

$$\bar{v} = \left(\frac{2kT}{m\pi}\right)^{1/2}$$

When the path length of the transition is called δ, the average lifetime in the transition state is

$$\tau = \frac{\delta}{\bar{v}} = \delta\left(\frac{m\pi}{2kT}\right)^{1/2}$$

As a result the concentration change per unit of time of the particles which are moving to the right is

$$\frac{(1/2)[XYZ^{\neq}]}{\tau} = \frac{(1/2)[XYZ^{\neq}]}{\delta(m\pi/2kT)^{1/2}} = \left(\frac{kT}{2m\pi}\right)^{1/2} \frac{[XYZ^{\neq}]}{\delta} \tag{4.23}$$

The total number of particles moving to the left must be equal to the number of particles moving to the right because the assumption was made that the system is in equilibrium. In reality, the motion to the left, that is, the direction of forming reactants, does not occur, but this has no influence on the calculation of the velocity of decomposition of the transition state.

The situation is comparable to a liquid-vapor surface. In the equilibrium state per unit of time as many particles move from liquid to vapor as from vapor to liquid. When the vapor is suddenly removed, the number of particles moving from liquid to vapor is still the same.

The rate of the reaction is

$$\left(\frac{kT}{2m\pi}\right)^{1/2} \frac{[XYZ^{\neq}]}{\delta}$$

and $[XYZ^{\neq}] = K[XY][Z]$, so that the rate is

$$\frac{K}{\delta}\left(\frac{kT}{2m\pi}\right)^{1/2} [XY][Z]$$

Empirically, the rate is given by $k_r[XY][Z]$, and thus

$$k_r = \frac{K}{\delta}\left(\frac{kT}{2m\pi}\right)^{1/2} \tag{4.24}$$

It will be recalled that this relation applies to both unimolecular and bimolecular processes (see page 70). The equilibrium constant K has been expressed by Eq. (4.12).

If the activated complex were a regular molecule rather than a molecule in a transition state, it would have a symmetric and an antisymmetric stretching vibration. Actually, for $XYZ^{\neq}$ the antisymmetric stretching vibration is replaced by a translational mode, $Q_{transl\ Y}$. We separate this special mode:

$\vec{X}$---Y---$\overset{\leftarrow}{Z}$ symmetric

$\overset{\leftarrow}{X}$---$\vec{Y}$---$\overset{\leftarrow}{Z}$ antisymmetric

X---$\vec{Y}$---Z translation

$$Q_{XYZ^{\neq}} = Q^{\neq} Q_{transl\ Y} \tag{4.25}$$

$Q^{\neq}$ involves the same partition functions as the total Q of a regular molecule XYZ but without the vibration which has become a translation. During the reaction all the degrees of freedom contributing to $Q^{\neq}$ are supposed to be in equilibrium with their surroundings.

How large is $Q_{transl\ Y}$? The path length of the transition is δ. Quantum mechanics tells us that the energy levels E_n of a particle with mass m in a one-dimensional box of length δ are

$$E_n = \frac{n^2h^2}{8m\delta^2}, \qquad n = 1, 2, 3, \ldots$$

$$Q_{transl\ Y} = \sum_{n=1}^{\infty} \exp \frac{-(n^2h^2)}{8m\delta^2kT}$$

If m is the mass of a carbon atom and if, furthermore, $\delta = 0.5$ nm and T = 300 K, then $h^2/(8m\delta^2kT) = 2.66 \times 10^{-3}$. As succeeding terms differ by very little, it is permissible to replace the sum by an integral (Fig. 4.6),

K = degree Kelvin

$$\int_0^{\infty} \exp\left[\frac{-(n^2h^2)}{8m\delta^2kT}\right] dn = \frac{(2\pi mkT)^{1/2}\delta}{h} \qquad \text{(cf. page 78)}$$

$$Q_{XYZ^{\neq}} = Q^{\neq} \frac{(2\pi mkT)^{1/2}\delta}{h} \tag{4.26}$$

which on substitution into Eqs. (4.12) and (4.24) gives

$$k_r = \frac{Q^{\neq}}{Q_{XY}Q_Z} \frac{(2\pi mkT)^{1/2}\delta}{h} \exp\left(\frac{-\Delta\varepsilon_0}{kT}\right)\frac{1}{\delta}\left(\frac{kT}{2m\pi}\right)^{1/2}$$

$$= \frac{kT}{h} \frac{Q^{\neq}}{Q_{XY}Q_Z} \exp \frac{-\Delta E_0}{RT} \tag{4.27}$$

if $\Delta E_0 = L\ \Delta\varepsilon_0$, where L is Avogadro's constant. The expressions for both the bimolecular and the monomolecular rate constants contain the factor kT/h. This factor is generally adopted for the equations of reactions of any order:

$$k_r = \frac{kT}{h} \frac{Q^{\neq}}{Q_A Q_B Q_C \cdots} \exp \frac{-\Delta E_0}{RT} \tag{4.28}$$

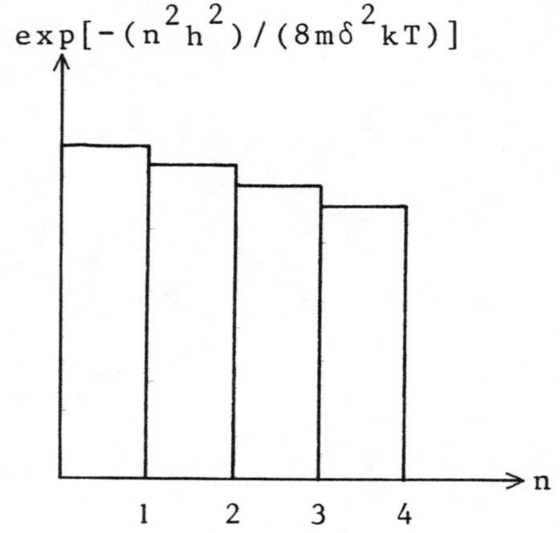

FIG. 4.6

A + B + C + ··· ⟶ products

PSEUDO-THERMODYNAMICAL EQUATIONS

It is possible now to calculate the rate constant theoretically when the partition function Q and ΔE_0 are known. For a few simple gas reactions it has been possible to calculate a rate constant which agrees roughly with the experimental one. Generally, in organic chemistry molecules are complicated and solvent effects are appreciable, and therefore these theoretical calculations cannot be realized. Yet the expression has considerable usefulness in analyzing reaction processes. Apart from the fact that one of the partition functions has been separated, the term

$$\frac{Q^{\neq}}{Q_{XY}Q_Z} \exp \frac{-\Delta E_0}{RT}$$

in Eq. (4.27) is equal to an equilibrium constant. We introduce the pseudo-equilibrium constant

$$K^{\neq} = \frac{Q^{\neq}}{Q_{XY}Q_Z} \exp \frac{-\Delta E_0}{RT} \qquad (4.29)$$

This converts Eq. (4.27) to

$$k_r = \frac{kT}{h} K^{\neq} \qquad (4.30)$$

At constant temperature and pressure an equilibrium constant K is related to the increase in standard molar Gibbs function $\Delta G_m^{\ominus}$ when 1 mol is transferred from one side to the other of the equilibrium:

$$K = \exp \frac{-\Delta G_m^{\phi}}{RT} \qquad (4.31)$$

Analogously, a similar expression has been introduced for $K^{\neq}$, and thus

$$k_r = \frac{kT}{h} \exp \frac{-\Delta G^{\neq}}{RT} \qquad (4.32)$$

$\Delta G^{\neq}$ is called the Gibbs function of activation. $\Delta G^{\neq}$ can be regarded as the difference in standard molar Gibbs function between the activated and initial states. For mixtures and solutions these Gibbs functions

$\Delta G^{\neq}$ is frequently called the free energy of activation and is sometimes symbolized as $\Delta F^{\neq}$.

are partial quantities, sometimes called the chemical potentials and denoted as $\mu_A^{\ominus}$ (for compound A). According to thermodynamics, $\Delta G = \Delta H - T\,\Delta S$. ΔH is the change in enthalpy, the heat effect, and ΔS is the change in entropy. Consequently,

$$k_r = \frac{kT}{h} \exp \frac{-\Delta H^{\neq}}{RT} \exp \frac{\Delta S^{\neq}}{R} \tag{4.33}$$

where $\Delta H^{\neq}$ and $\Delta S^{\neq}$ are the activation enthalpy and activation entropy, respectively. $\Delta G^{\neq}$, $\Delta H^{\neq}$, and $\Delta S^{\neq}$ are the basis for a quasi-thermodynamic treatment of reaction rates. For each k_r the $\Delta G^{\neq}$ can be calculated by applying Eq. (4.32). By measuring k_r at different temperatures we can obtain experimental values for $\Delta H^{\neq}$ and $\Delta S^{\neq}$:

$$\Delta G^{\neq} = -RT \ln\left(k_r \frac{h}{kT}\right)$$

$$\Delta H^{\neq} = RT^2 \frac{d \ln k_r}{dT} - RT \quad \text{and} \quad \frac{d \ln k_r}{dT} = \frac{1}{T} + \frac{\Delta H^{\neq}}{RT^2}$$

$$\Delta S^{\neq} = \frac{\Delta H^{\neq} - \Delta G^{\neq}}{T}$$

Graphically, $\Delta H^{\neq}$ can be obtained from a plot of ln(k/T) versus 1/T. The slope of such a plot is $-\Delta H^{\neq}/R$, Eq. (4.33). Most commonly, chemists plot ln k instead of ln(k/T), which gives the Arrhenius activation energy E_a [see Eq. (4.6)]. From the Arrhenius equation (4.6) and the equations for $\Delta H^{\neq}$ it follows that $E_a/RT^2 = \Delta H^{\neq}/RT^2 + 1/T$ or $E_a = \Delta H^{\neq} + RT$. The difference between the activation energy according to

$$k = A \exp \frac{-E_a}{RT}$$

Arrhenius, E_a, and the activation enthalpy according to transition-state theory, $\Delta H^{\neq}$, i.e., RT, is usually relatively small.

VOLUME OF ACTIVATION

For the relation between an equilibrium constant K and the pressure p at constant temperature T,

$$\left(\frac{d \ln K}{dp}\right)_T = \frac{-\Delta V_m}{RT}$$

where ΔV_m is the change of volume when 1 mol of compound is transferred from one side of the equilibrium to the other. Analogously, in reaction kinetics the expression

$$\left(\frac{d \ln k_r}{dp}\right)_T = \frac{-\Delta V^{\neq}}{RT}$$

is used, where $\Delta V^{\neq}$ is equal to $d(\Delta G^{\neq})/dp$. Thus, $\Delta V^{\neq}$ can be regarded as the molar volume change in going from the initial to the transition state. Just as in the case of other quasi-thermodynamic quantities, $\Delta V^{\neq}$ is used to draw conclusions about the reaction mechanism. How is $\Delta V^{\neq}$ connected with molecular changes? Suppose a molecule decomposes in passing from the initial state to the activated complex; that is, it is undergoing expansion on reaching the transition state. Thus, in the transition state of the slow step $\Delta V^{\neq} > 0$, and the reaction rate decreases with pressure. However, when two molecules are brought

$$A \longrightarrow a_1\text{----}a_2$$

$$a_1 + a_2 \longrightarrow A$$

together in the activated complex, usually $\Delta V^{\#} < 0$, and the reaction rate increases with pressure. These conclusions might be disturbed by solvation. Solvation changes obviously can cause volume changes.

To estimate the pressure necessary to influence reaction rate, the following calculation can be carried out. When two molecules combine, their van der Waals interaction at a distance of, e.g., 0.36 nm changes to a partial covalent bond with a length of, e.g., 0.16 nm. If the diameter of the molecule at the position of the bond is 0.10 nm^2, a value which is not unusual, then

$$\Delta V^{\#} = -(0.36 - 0.16) \times 0.10 = -0.020 \; \text{nm}^3/\text{molecule}$$

$$= -0.020 \times 10^{-21} \times 6 \times 10^{23} = -12 \; \text{cm}^3/\text{mol}$$

$$RT = 8.3 \times 300 \; \text{J/mol} = 2490 \; \text{Nm/mol}$$

$$\frac{d \ln k_r}{dp} = \frac{dk_r/dp}{k_r} = +48 \times 10^{-10} \; \text{m}^2/\text{N} \approx 48 \times 10^{-5} \; \text{atm}^{-1}$$

If, for instance, $\Delta k_r/k_r = 0.5$,

$$\Delta p = \frac{0.5}{48 \times 10^{-5}} = 1000 \; \text{atm}$$

That is, in this example a 1000-atm pressure change is required to produce a 50% rate change.

For a more exact calculation it would be necessary to integrate.

Special equipment is required for such high pressures.

$$\Delta V^{\#} = \Delta V_1^{\#} + \Delta V_2^{\#}$$
$\uparrow$ reacting compounds $\qquad$ $\uparrow$ solvation

E. Whalley, Advan. Phys. Org. Chem. 2:108 (1964).

$1 \; \text{nm} = 10^{-9} \; \text{m} = 10^{-7} \; \text{cm} = 10 \; \text{Å}$

$1 \; \text{J} = 1 \; \text{Nm} = 1 \; \text{newton} \times 1 \; \text{meter}$

$0.987 \; \text{atm} = 10^5 \; \text{N/m}^2$

The SI unit of pressure is the Pascal: $1 \; \text{Pa} = 1 \; \text{N/m}^2$.

W. J. le Noble, Progr. Phys. Org. Chem. 5:207 (1967).

K. R. Brower and J. S. Chen, J. Amer.
Chem. Soc. $\underline{87}$:3396 (1965).

Examples: For the solvolysis of tert-amyl chloride in 80% aqueous
ethanol at 34.2°C, the volume of activation appears to be
$\Delta V^{\neq} = -18 \text{ cm}^3/\text{mol}$.

The slow step is an ionization:

$$\begin{array}{c} | \\ -C-Cl \\ | \end{array} \longrightarrow \left[\begin{array}{c} | \\ -C^{\delta+} \cdots Cl^{\delta-} \\ | \end{array} \right]^{\neq} \longrightarrow \begin{array}{c} | \\ -C^{+} \\ | \end{array} \cdot Cl^{-}$$

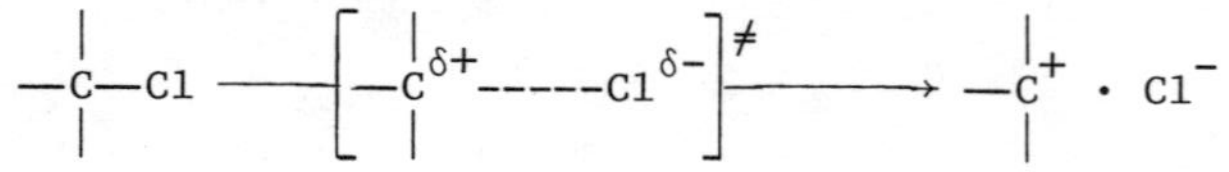

$\Delta V^{\neq}$ is negative, although the bond between carbon and chlorine is
stretched and the molar volume is increased. The reason for this discrepancy must be solvation. Solvation is of much greater importance in
the rather electrovalent transition state than in the covalent initial
state. The solvent molecules are more tightly packed in the solvation
shell of the ion pair developing in the transition state than in the
bulk of the solvent. This phenomenon is called, in general,
electrostriction.

For the solvolysis of tert-amyldimethylsulfonium iodide at 53.8°C
in the same solvent, $\Delta V^{\neq} = +14.3 \text{ cm}^3/\text{mol}$:

$$\begin{array}{c} C_2H_5 \quad CH_3 \\ | \quad\quad | \\ CH_3-C-S^+ \\ | \quad\quad | \\ CH_3 \quad CH_3 \end{array} \longrightarrow \left[\begin{array}{c} C_2H_5 \quad\quad CH_3 \\ | \quad\quad\quad\quad | \\ CH_3-C^{\delta+}\cdots S^{\delta+} \\ | \quad\quad\quad\quad | \\ CH_3 \quad\quad CH_3 \end{array} \right]^{\neq} \longrightarrow \begin{array}{c} C_2H_5 \\ | \\ CH_3-C^+ \\ | \\ CH_3 \end{array} + S(CH_3)_2$$

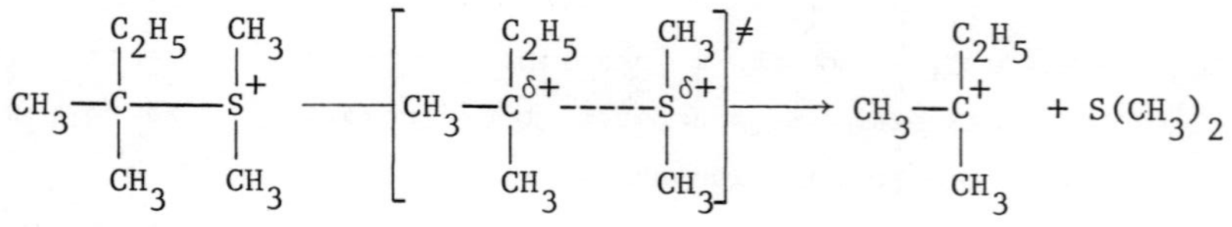

In this case the transition state involves diffusion of charge away from the charge center in the initial state. This results in destruction of solvent organization in the initial state and, therefore, the opposite of electrostriction in going to the transition state:

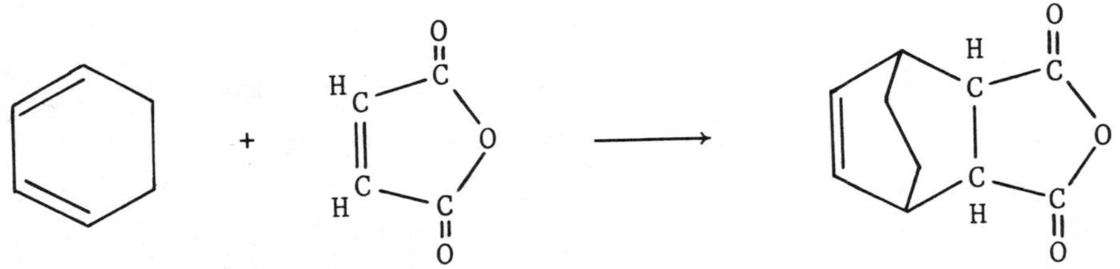

Diels-Alder reactions are characterized by highly negative $\Delta V^{\neq}$-values. For example, the reaction between 1,3-cyclohexadiene and maleic anhydride in dichloromethane has a $\Delta V^{\neq}$-value of -37.2 cm^3/mol. The appreciable volume change indicates that in the transition state the bonds to appear in the product have been formed to a large extent:

$$C_6H_5N(CH_3)_2 + C_2H_5I \longrightarrow C_6H_5\overset{+}{N}(CH_3)_2(C_2H_5) \cdot I^-$$

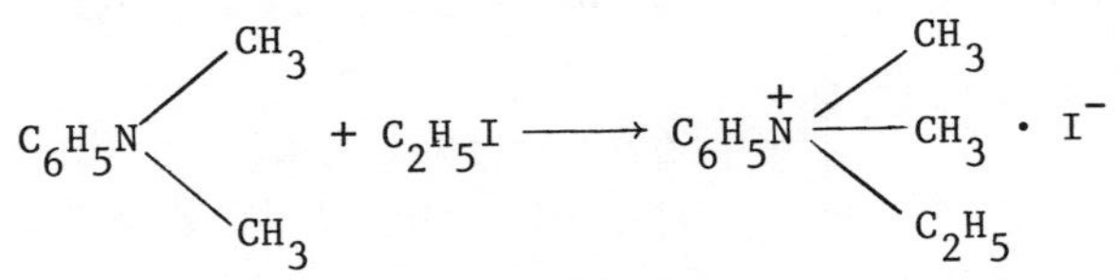

This quaternization reaction has $\Delta V^{\neq} = -20$ cm^3/mol when performed in nitrobenzene and $\Delta V^{\neq} = -34$ cm^3/mol in methanol. The difference is due to the ion-pair characteristic developed in the transition state. Here the charge centers are better solvated by methanol than by nitrobenzene.

R. A. Grieger and C. A. Eckert, J. Amer. Chem. Soc. 92:7149 (1970).

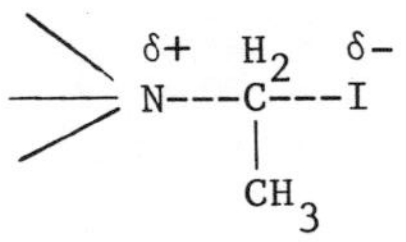

ENTROPY OF ACTIVATION

Instead of the activation volume, more often the activation entropy $\Delta S^{\neq}$ is used. The entropy of a system is related to its mobility and flexibility. A highly mobile and flexible system is associated with a large entropy term and a rigid system with a small one.

An example of a reaction in which two freely moving initial-state molecules are joined in the activated complex is the Diels-Alder reaction. Because of the imposed restriction of movement, $\Delta S^{\neq}$ is highly negative; the values vary between -100 and -20 J mol^{-1} K^{-1}. For the reverse reaction, the retro-Diels-Alder reaction, $\Delta S^{\neq}$ is much larger: between -20 and +70 J mol^{-1} K^{-1}.

H. Kwart and K. King, Chem. Rev. 68:436 (1968).

Frequently, the activation entropy is negative because formation of a transition state often involves freezing out of some rotations. In particular, the alterations of rotational and translational modes contribute to the entropy change more than the vibrational. This is due to the fact that in general the entropy change from initial to transition state is affected by the components of the partition function in the sequence:

$$Q_{transl} >> Q_{rot} > Q_{vibr}$$

The enthalpy change is controlled largely by the components of the partition function in the sequence $Q_{transl} >> Q_{rot} > Q_{vibr}$

L. L. Schaleger and F. A. Long, Advan. Phys. Org. Chem. 1:1 (1963).

Example: The thermolysis reaction mentioned on page 53 has a cyclic transition state, $\Delta S^{\neq} = -63$ J mol^{-1} K^{-1}, or -15 e.u.

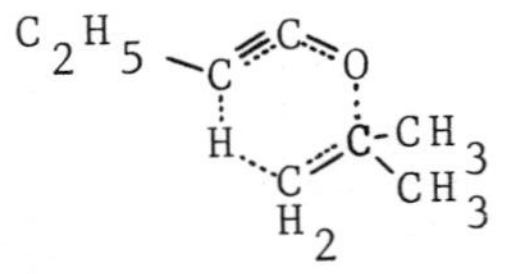

Until now, generally, $\Delta S^{\neq}$ has been expressed in units of cal mol^{-1} $degree^{-1}$, abbreviated as e.u. (entropy unit). The SI units are J mol^{-1} K^{-1}.

1 e.u. = 4.186 J mol^{-1} K^{-1}

TRANSMISSION COEFFICIENT

For the sake of completeness we must also mention the so-called transmission coefficient κ:

$$k_r = \kappa \frac{kT}{h} e^{-\Delta G^{\neq}/RT}$$

In the theoretical expression for the rate constant we have not yet reckoned with the possibility that the activated complex may return partially to the initial reactants. If this happens, the actual rate constant is less than the calculated one. The correction can be applied by giving the transmission coefficient a value smaller than unity. For some gas reactions the reverse reaction is considerable. Assume that two particles react with each other in the gas phase with liberation of energy. The system AB wants to get rid of this energy, and this can occur by way of collision with a third particle. In a very dilute gas phase the time between two collisions can be so long that the energy-rich complex has time to dissociate into the reactants A and B. In organic chemistry, reactions in very dilute gas phase are of lesser importance. Therefore, in general it is not often necessary to consider a value of $\kappa < 1$.

$$A + B \longrightarrow C$$

TUNNEL EFFECT

However, κ can be greater than unity because of the tunnel effect. Normally, a system reacts only if it has sufficient energy to reach the transition-state mountain top. See Fig. 4.7. Systems with an energy surplus < $\Delta G^{\neq}$ with respect to the initial state cannot normally be expected to react. According to quantum mechanics, however, this view is not completely correct. It is not at all certain that a system passing along the reaction coordinate remains to the left of the energy barrier to reaction when its energy surplus is smaller than $\Delta G^{\neq}$. If the energy and momentum can be exactly fixed, then the location of the particle on the energy surface is associated with a degree of uncertainty stipulated by the Heisenberg Uncertainty Principle. The particle may also be found on the other side of the reaction barrier. If at any point along the reaction coordinate to the left or right of the transition-state maximum the uncertainty in the particle's location is greater than the barrier thickness, the particle may be said to have "tunneled" through the energy mountain without ever having realized the $\Delta G^{\neq}$ for going over the mountain top. On the other hand, molecules with an energy larger than $\Delta G^{\neq}$ can partially reverse in their course, tending to lower the value of κ, but the total effect is an increase in rate and leads to $\kappa > 1$.

Although in most cases this effect can be neglected, it is sometimes observed in reactions involving hydrogen or electron transfer. These lighter particles have a greater chance to deviate from classical behavior. The tunnel effect is even smaller for the heavier deuterium

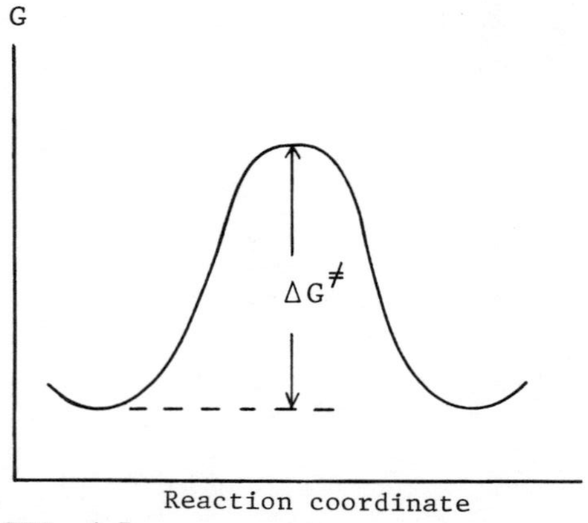

FIG. 4.7

Heisenberg Uncertainty Principle:

$\Delta p \Delta q \leq h/2\pi$

$\Delta E \Delta t \leq h/2\pi$

p = momentum
q = location
E = energy
t = lifetime
h = Planck's constant

than for hydrogen. These considerations will be treated at greater length in Chap. 5.

COMPENSATION LAW AND THE ISOKINETIC RELATIONSHIP

The reaction rate increases when $\Delta H^{\neq}$ becomes smaller or when $\Delta S^{\neq}$ becomes larger. Often an increase in reaction rate owing to a smaller $\Delta H^{\neq}$ tends to be partially compensated by a simultaneous decrease of the reaction rate owing to a smaller $\Delta S^{\neq}$; the reverse situation may also occur. It is quite common for the overall rate change to be smaller than could be expected from the change in $\Delta H^{\neq}$ or in $\Delta S^{\neq}$ separately. This phenomenon is called the "compensation law," although it is more of a "tendency" than a law.

$$k_r = \kappa \frac{kT}{h} \exp \frac{-\Delta H^{\neq}}{RT} \exp \frac{\Delta S^{\neq}}{R}$$

J. E. Leffler, J. Org. Chem. 20:1202 (1955).

The cause of this compensation is not always clear and attributable to a single cause. Sometimes solvation factors may be at the origin of this tendency. For example, suppose that owing to a change of solvent the extent of solvation of the transition state increases. Stabilization of the activated complex through increased solvation causes a decrease in $\Delta H^{\neq}$. However, the greater degree of solvation also means "freezing" of the solvent molecules, that is, a greater degree of organization in solvent structure of the transition state and a consequent decline in $\Delta S^{\neq}$.

These changes in $\Delta H^{\neq}$ and $\Delta S^{\neq}$, $\Delta(\Delta H^{\neq})$ and $\Delta(\Delta S^{\neq})$, respectively, can occur through a change in solvent nature or through a change in substitution, near the reaction center. Sometimes a linear relation is observed between $\Delta H^{\neq}$ and $\Delta S^{\neq}$ (see Fig. 4.8):

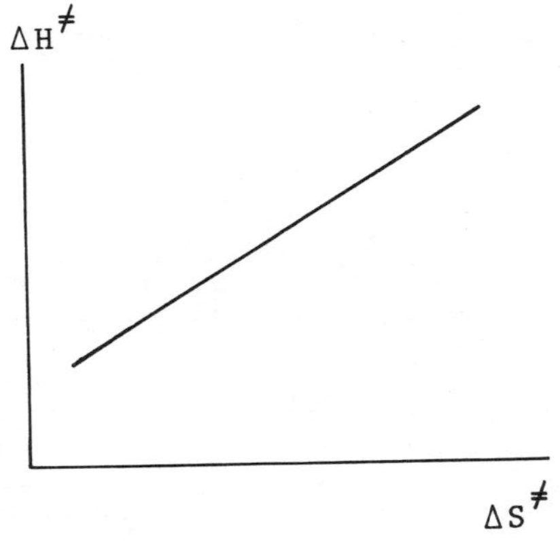

FIG. 4.8

$$\Delta(\Delta H^{\neq}) = \beta\ \Delta(\Delta S^{\neq}), \qquad \beta > 0$$

Insertion in $\Delta(\Delta G^{\neq}) = \Delta(\Delta H^{\neq}) - T\ \Delta(\Delta S^{\neq})$ gives

$$\Delta(\Delta G^{\neq}) = \Delta(\Delta H^{\neq})\left(1 - \frac{T}{\beta}\right) \tag{4.34}$$

This is the so-called isokinetic relationship. $\Delta(\Delta G^{\neq}) = 0$ when $T = \beta$, or, at this temperature, called the isokinetic temperature, the structural or solvent changes under consideration affect neither $\Delta G^{\neq}$ nor the rate of reaction. An example is the internal 1,3-dipolar addition of 2-azidobenzophenones with various substituents R. At the isokinetic temperature of 160°C the substituent effect essentially disappears.

J. E. Leffler and E. Grunwald, Rates and Equilibria of Organic Reactions (Wiley, New York, 1963), p. 324.

J. Hall, F. E. Behr, and R. L. Reed, J. Amer. Chem. Soc. 94:4952 (1972).

R. C. Petersen, J. Org. Chem. 29:3133 (1964).
Statistical treatment: O. Exner, Progr. Phys. Org. Chem. 10:411 (1973).

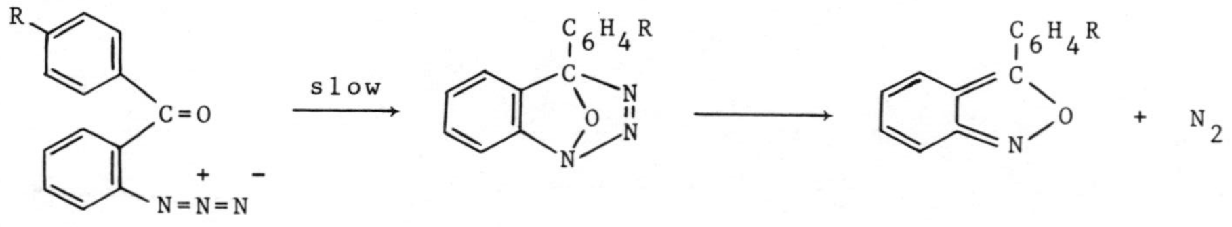

One has to keep in mind that $\Delta H^{\neq}$ and $\Delta S^{\neq}$ are both derived from the same series of experimental data. Errors in these data appear in $\Delta H^{\neq}$ as well as in $\Delta S^{\neq}$, and variations in $\Delta H^{\neq}$ and $\Delta S^{\neq}$ are not independent of each other. Thus, many published values of β may be artifacts.

ACTIVITY COEFFICIENTS

On page 71 we introduced the equilibrium constant K which describes the relation between the concentration of the activated complex and the concentration of reactant in the initial state. For $XY + Z \rightleftharpoons XYZ^{\neq}$

$$K = \frac{[XYZ^{\neq}]}{[XY][Z]} \qquad \text{and} \qquad [XYZ^{\neq}] = K[XY][Z] \tag{4.35}$$

If the activity coefficients of the particles are not unity, then, according to thermodynamics,

$$K = \frac{[XYZ^{\neq}]}{[XY][Z]} \frac{f_{XYZ^{\neq}}}{f_{XY} f_Z} \tag{4.36}$$

activity = f × concentration

The f's are the activity coefficients.

The rate of reaction still depends on the concentration of the complexes $XYZ^{\neq}$. For this concentration, instead of Eq. (4.35) we now have

$$[XYZ^{\neq}] = K \frac{f_{XY} f_Z}{f_{XYZ^{\neq}}} [XY][Z] \tag{4.37}$$

Accordingly, in the equations for the rate constant the equilibrium constants K and $K^{\neq}$, as they appear in Eqs. (4.24) and (4.30), are modified by the activity coefficient ratios and become

$$K \frac{f_{XY} f_Z}{f_{XYZ^{\neq}}} \qquad \text{and} \qquad K^{\neq} \frac{f_{XY} f_Z}{f_{XYZ^{\neq}}}$$

respectively.

Remark: The $f_{XYZ^{\neq}}$ in the equation for $K^{\neq}$ does not involve the translational mode (antisymmetric vibration); see page 80. Equation (4.30) for the rate constant thus is transformed into

$\overleftarrow{X} \ldots \overrightarrow{Y} \ldots \overleftarrow{Z}$

$$k_r = \frac{kT}{h} K^{\neq} \frac{f_{XY} f_Z}{f_{XYZ^{\neq}}} \qquad \text{or} \qquad k_r = (k_r)_0 \frac{f_{XY} f_Z}{f_{XYZ^{\neq}}} \tag{4.38}$$

The latter equation is also called Brønsted's rate equation. The rate constant reduces to $(k_r)_0$ under circumstances where all activity coefficients are equal to 1.

MICROSCOPIC REVERSIBILITY

The system $A \rightleftarrows B$ is tending toward an equilibrium. When we start with only compound A, the reaction takes place via the energy barriers proceeding from left to right. When we start with only compound B, the same energy path going in reverse is followed in precise detail from right to left, even when only little of B is transformed to A and the equilibrium is largely displaced to the right. The mechanism for the reaction from right to left follows precisely the same reaction coordinate but in the reverse direction. This is called the principle of microscopic reversibility (Fig. 4.9).

$A \rightleftarrows B$

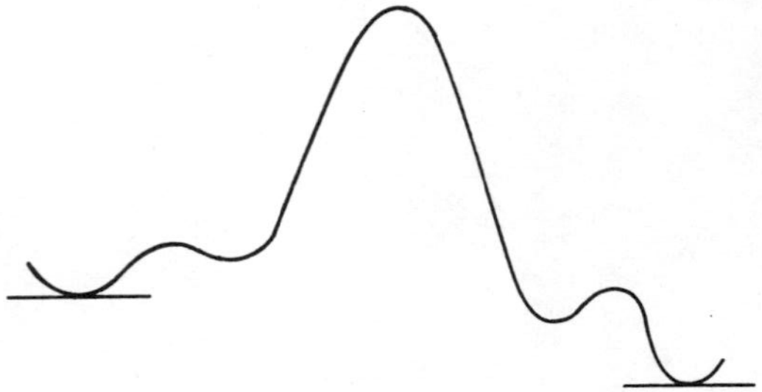

A B

FIG. 4.9

Example: Suppose a radical X˙ leaves a molecule

```
 |   |
XC—CY
 |   |
```

only when X and Y are in antiperiplanar, app, positions.
Then, according to the principle of microscopic reversibility, in the

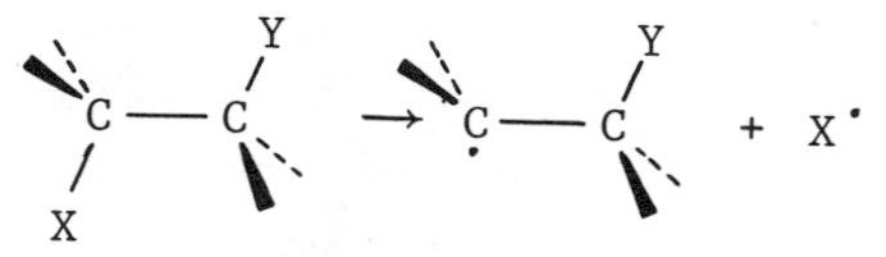

reverse (addition) reaction X˙ will be added in the same app relationship with respect to Y.

The application of this principle and the rules governing reaction mechanisms which it engenders are often of great benefit in studies of the course of chemical reactions. For example, where the forward reaction does not readily lend itself to kinetic observation, it is frequently possible to study the reverse pathway with the understanding that the mechanistic features of the reaction in the direction of interest are also being elucidated. Typical application of this principle: Substituent and isotope effects indicate that the rearrangement

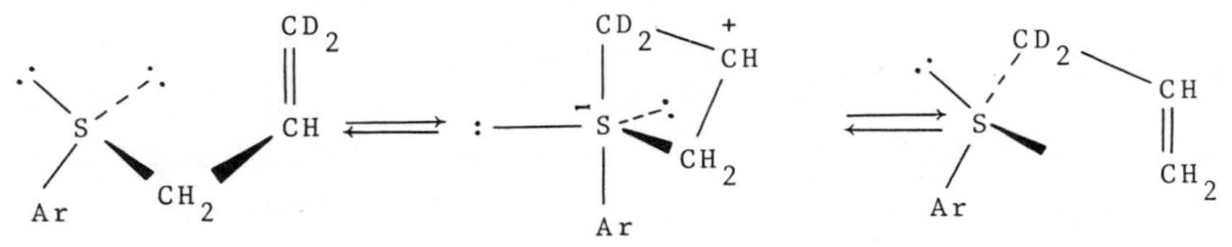

1 2 3

H. Kwart and N. A. Johnson, J. Amer. Chem. Soc. 99:3441 (1977).

1 → 3 passes through an intermediate which could have the trigonal bipyramidal structure 2. The vertical bonds in 2 are called apical; they are distinctly different (and weaker) than the three bonds in the horizontal plane which are called basal. In the formation of the four-membered ring (1 → 2) an apical bond is formed. The principle of microscopic reversibility requires that the opening of this ring must take place with apical bond breaking. It is necessary, therefore, that before 2 rearranges to 3 an energy barrier must be surmounted; i.e., a rearrangement of bonds must be experienced (sometimes called a polytopal rearrangement or pseudo-rotation) converting 2 into the structure 2b in which the —CH_2— has become apical and the CD_2 basal.

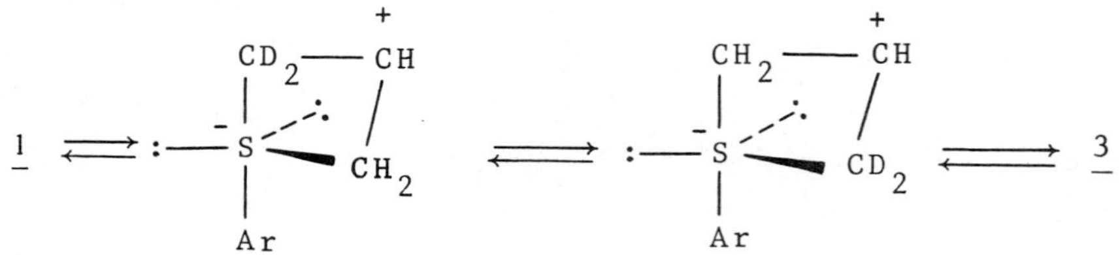

2 2b

One way this could occur, for example, would be via the more energetic square pyramid intermediate 2a in which the CH_2 and CD_2 bonds to sulfur were of identical energy.

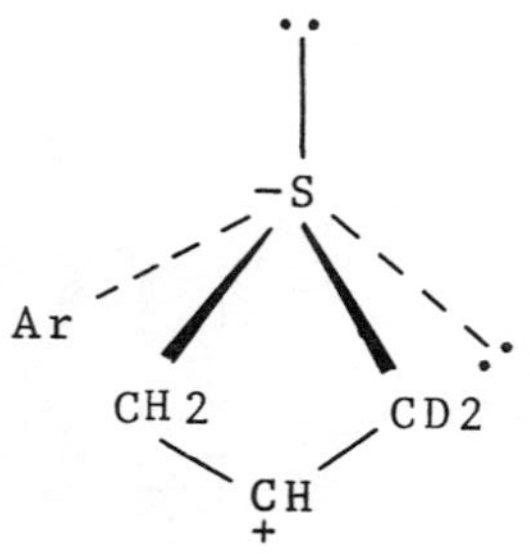

2a

In another way of looking at this case we may suppose that the reaction 1 → 2 → 3, where also the deuterium has been replaced by hydrogen, would occur without intermediate 2b being involved at all. The energies of $\underline{1}$ and $\underline{3}$ are identical since the products and reactants are identical. Consequently the formation of the apical bond in the step $\underline{1} \rightarrow \underline{2}$ does not fully compensate for the breaking of the basal bond in step $\underline{2} \rightarrow \underline{3}$ because these bonds are distinctly different in energy. Thus, the overall process $\underline{1} \rightarrow \underline{3}$ and its reverse would either be creating or destroying energy in violation of the second law. Clearly, then, the principle of microscopic reversibility is a derivative of the second law of thermodynamics.

The two-dimensional reaction coordinate diagram depicting the transformations of ground and transition states has been extremely useful in the analysis of reactivities in organic chemistry. Consider the following illustrative system

$$P_A \xleftarrow{k_A} A \underset{k_{-1}}{\overset{k_1}{\rightleftarrows}} B \xrightarrow{k_B} P_B$$

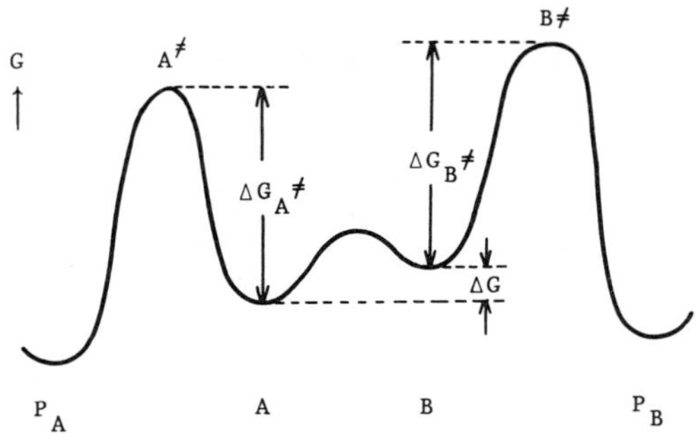

where the equilibration $A \rightleftarrows B$ is rapid compared with formation of the products P_A and P_B. The equilibrium constant $[A]/[B] = k_{-1}/k_1 = K$.

For instance, A and B could be two rapidly interconverting conformations of one compound. The overall rate constant for the reaction of this compound is given by the expression

$$k = k_A[A] + k_B[B]$$

If k_A and k_B are known or can be estimated from appropriate model compounds, the position of the conformational equilibrium [A]:[B] can be calculated from the overall rate constant k. This is known as the

Winstein-Holness analysis, named after the authors who first devised and applied the treatment. They measured rates of reactions of cyclohexyl compounds 4-R-C_6H_{10}-X in order to arrive at the conformational composition expressed as the $X_{axial}:X_{equatorial}$ ratio. The derivatives with R = tert-butyl were used as the rate reference compounds:

S. Winstein and N. J. Holness, J. Amer. Chem. Soc. 77:5562 (1955).

trans-4-tert-butyl	for $k(X_{equatorial})$
cis-4-tert-butyl	for $k(X_{axial})$

Another way of analyzing such a system is to determine the product ratio P_A/P_B. The rate of formation of P_A is $d[P_A]/dt = k_A[A]$, and of P_B it is $d[P_B]/dt = k_B[B]$. Since the equilibration is rapid the ratio [A]/[B] is a constant during the whole process and, therefore, $P_A/P_B = k_A[A]/k_B[B] = (k_A/k_B)([A]/[B])$. Since k_A is determined by $\Delta G_A^{\neq}$, k_B by $\Delta G_B^{\neq}$ and [A]/[B] by ΔG, it can easily be seen that the ratio $[P_A]/[P_B]$ is exclusively determined by the difference in Gibbs function between the transition states $A^{\neq}$ and $B^{\neq}$ and not by the initial states A and B. This conclusion is known as the Curtin-Hammett principle.

L. P. Hammet, Phys. Org. Chem., 2nd ed., p. 119 (McGraw-Hill, New York, 1970).

J. I. Seeman and W. A. Farone have calculated exact solutions for this system

J. Org. Chem. 43:1854 (1978).

$$P_A \xleftarrow{k_A} A \underset{k_{-1}}{\overset{k_1}{\rightleftarrows}} B \xrightarrow{k_B} P_B$$

regardless of the relative values of the rate constants, as long as the reactions are first or pseudo-first order. A major conclusion derived from their treatment is that, in general, the Curtin-Hammett principle and the Winstein-Holness equation correctly approximate the exact solution when k_1 and k_{-1} are at least ten times as large as k_A and k_B.

5 Isotope Effects

More decisive mechanistic criterion than substituent-rate effects

One of the most widely applied approaches to the study of organic reaction mechanisms is based on the *substituent effect*. This procedure involves modification of the reactant structure through introduction of a suitable substituent and observing the effect of this change on reaction rate and/or product distribution. Although the classical substituent effect is usually of large magnitude, interpretation is rarely unequivocal.

In general, we speak of a reaction series comprised of reactants differing only in the nature of a substituent located at a fixed position with respect to the seat of reaction common to all the series members. An example is the series of *para* substituted benzyl chlorides, where the substituent may be varied from p-methoxy to p-nitro at both

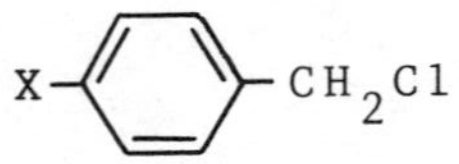

ends of the polarity scale. The series may be examined with regard to the substituent influence on the rate of solvolysis as a means of elucidating the fundamental mechanistic features of the solvolytic process.

The substitution of an isotopic atom alters the properties of the parent molecule in the same general way as does the introduction of any other substituent. However, in general, structural substitutions result in electronic changes and their attendant consequences for reactivity, whereas in all cases of isotopic substitution the electron distribution in the original bond has been altered to only a negligible extent. The isotopic mass is a factor of far greater importance to the vibrational energy of a bond than to its electronic energy. Yet some properties such as rates and equilibria may be appreciably changed. The isotope effects on rates can be directly related to the reaction mechanism, and therefore they are a valuable tool in the study of reaction mechanisms. We start with <u>isotope effects for heavy atoms, i.e., other than hydrogen</u>, and we want to introduce the subject by first considering <u>equilibrium isotope effects</u>.

EQUILIBRIUM ISOTOPE EFFECT

As expressed in Eq. (4.10), the equilibrium constant for the reversible, unimolecular rearrangement

$$R \underset{}{\overset{K}{\rightleftarrows}} P \qquad\qquad R = \text{reactant},\quad P = \text{product}$$

is equal to

is equal to

$$K = \frac{\Pi\, Q_P}{\Pi\, Q_R} \tag{5.1}$$

where $\Pi\, Q$ designates the product of the various contributions to the total partition function. Substitution of an isotopic atom in P and in R leads to an identical equation for the substituted molecules. The isotope effect on the equilibrium is defined as

$$\frac{K_\ell}{K_h} = \frac{(\Pi\, Q_P/\Pi\, Q_R)_\ell}{(\Pi\, Q_P/\Pi\, Q_R)_h} \tag{5.2}$$

where the subscript ℓ indicates the light isotope and h the heavy. Bigeleisen and Mayer determined the partition function ratio. Since the electronic partition function is practically isotope-invariant, it is deleted. It is also assumed that translational, vibrational, and rotational interactions are negligible; i.e., their contributions to the total partition function are independent of each other. The Q_{transl} contribution to the isotope effect is

J. Bigeleisen and M. G. Mayer, J. Chem. Phys. 15:261 (1947).

$$\frac{(M_\ell/M_h)_P^{3/2}}{(M_\ell/M_h)_R^{3/2}} \tag{5.3}$$

M = molecular weight

since for each of the three translational dimensions Q is proportional to $(\text{mass})^{1/2}$ [cf. Eqs. (4.25) and (4.26)]. Similarly, the rotational contribution is equal to

$$\frac{(I_{x,\ell}I_{y,\ell}I_{z,\ell}/I_{x,h}I_{y,h}I_{z,h})_P^{1/2}}{(I_{x,\ell}I_{y,\ell}I_{z,\ell}/I_{x,h}I_{y,h}I_{z,h})_R^{1/2}} \tag{5.4}$$

where the I's are the moments of inertia with respect to rotation around the axes x, y, and z. The vibrational contribution to the isotope effect, neglecting anharmonicity corrections, is [cf. Eq. (4.17)]

$$\frac{(Q_{vibr,\ell}/Q_{vibr,h})_P}{(Q_{vibr,\ell}/Q_{vibr,h})_R} \tag{5.5}$$

Each

$$(Q_{vibr})_P = \prod_i (Q_{vibr,i})_P$$

Π = product

where the product is over all molecular vibrations (i) of the product molecule P with all degenerate levels counted the appropriate number of times. Similarly, each

$$(Q_{vibr})_R = \prod_j (Q_{vibr,j})_R$$

The isotope effect is given by the product of the expressions (5.3), (5.4), and (5.5). This product can be converted to a simpler expression by application of the so-called product rule, according to which

O. Redlich, Z. Phys. Chem., B 28:371 (1935).

The derivation of this rule is too complicated to be repeated in this limited discussion.

$$\left(\frac{M_\ell}{M_h}\right)^{3/2}\left(\frac{I_{x,\ell}I_{y,\ell}I_{z,\ell}}{I_{x,h}I_{y,h}I_{z,h}}\right)^{1/2} = \left(\frac{m_\ell}{m_h}\right)^{3/2} \prod_i \frac{u_{i,\ell}}{u_{i,h}} \tag{5.6}$$

where m_ℓ and m_h are the masses of the light and heavy isotopes and $u_i \equiv h\nu_i/kT$. Equation (5.6) applies to R as well as to P. Now, for K_ℓ/K_h we obtain the simpler expression

$$\frac{\left[\left(\frac{m_\ell}{m_h}\right)^{3/2} \prod_i \frac{u_{i,\ell}}{u_{i,h}} \frac{Q_{vibr,i,\ell}}{Q_{vibr,i,h}}\right]_P}{\left[\left(\frac{m_\ell}{m_h}\right)^{3/2} \prod_j \frac{u_{j,\ell}}{u_{j,h}} \frac{Q_{vibr,j,\ell}}{Q_{vibr,j,h}}\right]_R}$$

The m_ℓ/m_h ratio is the same in P and R and cancels. $Q_{vibr,i}$ is given by Eq. (4.17), and thus

$$\frac{K_\ell}{K_h} = \frac{\left[\prod_i \frac{u_{i,\ell}}{u_{i,h}} \frac{\exp(-u_{i,\ell}/2)}{1 - \exp(-u_{i,\ell})} \frac{1 - \exp(-u_{i,h})}{\exp(-u_{i,h}/2)}\right]_P}{\left[\prod_j \frac{u_{j,\ell}}{u_{j,h}} \frac{\exp(-u_{j,\ell}/2)}{1 - \exp(-u_{j,\ell})} \frac{1 - \exp(-u_{j,h})}{\exp(-u_{j,h})}\right]_R} \tag{5.7}$$

If we define $\Delta u_i = u_{i,\ell} - u_{i,h}$ and make the justifiable assumption that Δu_i is small, each of the bracketed terms in (5.7) can be rearranged to

$$1 - \sum_{i=1}^{3n-6} G(u_{i,h})\, \Delta u_i$$

where

$$G(u_{i,h}) = \frac{1}{2} - \frac{1}{u_{i,h}} + \frac{1}{\exp(u_{i,h}) - 1} \tag{5.8}$$

because, generally, $\exp(x) = 1 + x + x^2/2 + \cdots$ and for $x \ll 1$ $\exp(x) \simeq 1 + x$.

In linear molecules

$$\sum_{i=1}^{3n-5}$$

J. Bigeleisen and M. G. Mayer, J. Chem. Phys. 15:261 (1947), give a list of G-values for u = 0 to 25.

$$\frac{K_\ell}{K_h} = \frac{\left[1 - \sum_{i=1}^{3n-6} G(u_{i,h})\,\Delta u_i\right]_P}{\left[1 - \sum_{j=1}^{3n-6} G(u_{j,h})\,\Delta u_j\right]_R}$$

Since the numerator and denominator are both close to 1, we can rearrange this expression to arrive at

$$\frac{K_\ell}{K_h} = 1 - \left[\sum_{i=1}^{3n-6} G(u_{i,h})\,\Delta u_i\right]_P + \left[\sum_{j=1}^{3n-6} G(u_{j,h})\,\Delta u_j\right]_R \qquad (5.9)$$

Since the Gibbs function $\Delta G^{\ominus} = -RT \ln K$ [cf. Eq. (4.31)], the change in Gibbs function because of isotopic substitution is

$$\Delta G_\ell^{\ominus} - \Delta G_h^{\ominus} \equiv \Delta(\Delta G^{\ominus}) = RT \ln \frac{K_\ell}{K_h} \qquad (5.10)$$

KINETIC ISOTOPE EFFECT

The theory of the isotope effects on rates of reactions involves a fusion of Eyring's absolute reaction rate theory into the thermodynamic isotope effect treatment discussed above. As a result it is subject to the limitations of both. In the absolute rate theory, the reactants and products are considered to be in thermal equilibrium with an activated complex $A^{\neq}$:

$$R \rightleftarrows A^{\neq} \rightleftarrows P$$

$$\prod_{i=1}^{3n-6} \frac{\exp(-u_{i,\ell}/2)}{\exp(-u_{i,h}/2)} \simeq 1 - \sum_{i=1}^{3n-6} \frac{\Delta u_i}{2}$$

$$\prod_{i=1}^{3n-6} \frac{1 - \exp(-u_{i,h})}{1 - \exp(-u_{i,\ell})} \simeq 1 - \sum_{i=1}^{3n-6} \frac{\Delta u_i}{\exp(u_{i,h}) - 1}$$

$$\prod_{i=1}^{3n-6} \frac{u_{i,\ell}}{u_{i,h}} = 1 + \sum_{i=1}^{3n-6} \frac{\Delta u_i}{u_{i,h}}$$

The rate constant for the forward reaction is [cf. Eq. (4.24)]

$$k_r = \frac{K}{\delta}\left(\frac{kT}{2\pi m}\right)^{1/2} \tag{5.11}$$

where m is the mass of the particle moving along the reaction coordinate. Sometimes the mass cannot be assigned to a single moving particle but is related to aggregates of masses. In this case the term "effective mass," m*, is used. The rate constant for the reaction of R containing an isotopic atom shows an analogous dependence. The rate constant ratio or kinetic isotope effect is then given by [cf. Eq. (5.9)]

$$\frac{k_{r,\ell}}{k_{r,h}} = \left(\frac{m_h^*}{m_\ell^*}\right)^{1/2} \Bigg[1 + \underbrace{\sum_i G(u_{i,h})\,\Delta u_i}_{\text{term A for reactant}} - \underbrace{\sum_j G(u_{j,h}^{\neq})\,\Delta u_j^{\neq}}_{\text{term B for act. complex}}\Bigg] \tag{5.12}$$

This term $(m_h^*/m_\ell^*)^{1/2}$ is called the effective mass ratio. These summations are over all vibrations characteristic of the particular structure.

At very high temperatures the terms A and B tend to become vanishingly small because the u_i- and Δu_i-values are approaching zero. The bracketed value goes to unity, and Eq. (5.12) reduces to

$$\frac{k_{r,\ell}}{k_{r,h}} = \left(\frac{m_h^*}{m_\ell^*}\right)^{1/2} \tag{5.13}$$

i.e., a difference in reaction rates based solely on the effective masses of the atoms involved in motion along the reactive coordinate.

It is appropriate to mention at this point that until now we have been considering only transition states in which a critical bond is undergoing rupture (see page 80). However, there are numerous cases in which a new bond or a greater extent of bonding is occurring in the transition as compared to the initial state prior to any bond breaking.

At lower temperatures, the isotope effect is determined by the difference in value of terms A and B and by the term $(m_h^*/m_\ell^*)^{1/2}$. When the extent of bonding, i.e., coordination number of the reaction center in the activated complex, is identical to the bonding in the reactant, the two terms A and B in Eq. (5.12) cancel, and the isotope effect is equal to $(m_h^*/m_\ell^*)^{1/2}$, which is always greater than unity. When the degree of bonding is greater in the reactant compared to the activated complex, the bracketed term of (5.12) becomes greater than unity. On the other hand, more bonding or higher-frequency bonding in the transition state increases the value of term B versus term A in Eq. (5.12), and the bracketed term diminishes.

Summarizing, then, the isotope effect tends toward a maximum when term B has become very small, i.e., when the zero-point energies in the transition state are very low. On the other hand, k_ℓ/k_h tends to a minimum value, which may be < 1, where the degree of bonding to the isotopic center is greater in the activated complex than in the initial state. When the degree of bonding to the isotopic atom is nearly identical in initial and transition states, k_ℓ/k_h tends to an intermediate value that is greater than unity. Such a situation occurs, e.g., when the transition state can be regarded as concerted.

Since the structure and properties of a transition state are inaccessible to any direct, nonkinetic measurement, calculations of kinetic isotope effects require consideration of a reasonable model. Agreement between the calculated and experimental isotope effects may not guarantee that the assumed model is correct. At best, lack of agreement certainly proves the model wrong.

SIMPLIFIED METHOD FOR CALCULATION OF KINETIC ISOTOPE EFFECTS

As pointed out earlier, the magnitude of the kinetic isotope effect k_ℓ/k_h is given by the Bigeleisen equation (5.12). The exact solution of this equation for complex molecules, once a very cumbersome task, may now be accomplished with the aid of a computer and a proper computer program. It requires that all the vibrational frequencies be known for both the ground-state reactant and the proposed transition-state model. But, fortunately, several simplifying assumptions may be applied to the Bigeleisen equation to enable a ready, hand calculation of the isotope effect.

Equation (5.12) consists of two terms which are usually evaluated separately. The bracketed term is referred to as the temperature-dependent factor (TDF) and the remainder as the temperature-independent factor (TIF).

The calculation of the $^{32}S/^{34}S$ isotope effect for the thiaallylic rearrangement of α-methylallyl phenyl sulfide will serve to illustrate this simplified approach. Three types of processes which may be considered are a dissociative process with a transition state resembling an intimate ion pair, <u>1</u>; a concerted process where bond making and breaking

H. Kwart and J. Stanulonis, J. Amer. Chem. Soc. <u>98</u>:4009 (1976).

have occurred to the same extent in the transition state, modeled by 2; and an associative process where the sulfur seat of reaction has experienced an expansion of its valence due to bonding with the neighboring olefinic group, as represented in the transition-state model 3.

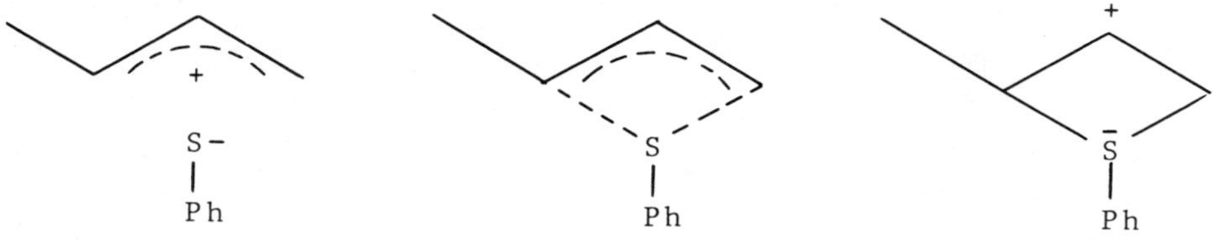

1 dissociative

2 concerted

3 associative

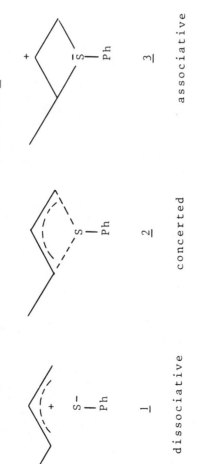

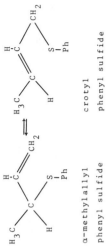

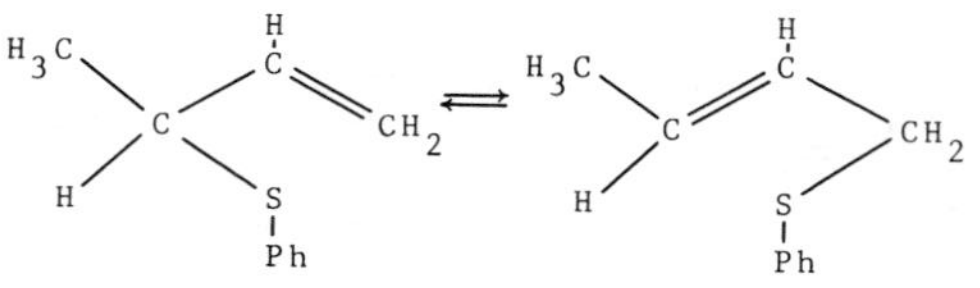

α-methylallyl phenyl sulfide

crotyl phenyl sulfide

The TIF is arrived at by focusing on the bond in the reactant which is to be broken and considering the masses on both sides of this center of bond breaking. For the dissociative model 1, the TIF is given as the square root of the ratio of the mass of the fragment containing the heavy isotope to the mass of the fragment containing the light isotope [Eq. (5.12)]; i.e.,

$$\mathrm{TIF} = \left(\frac{\text{mass Ph}^{34}\text{S}}{\text{mass Ph}^{32}\text{S}}\right)^{1/2} = \left(\frac{106}{104}\right)^{1/2} = 1.009$$

Since in the transition state modeled on 1 the bond to the isotopic atom has been broken, the TDF term reflects only the ground-state frequencies and according to Eq. (5.12) thus reduces to

$$\mathrm{TDF} = 1 + \sum_i G(u_i)\ \Delta u_i$$

In evaluating this term, furthermore, it is also assumed that only the symmetrical stretching vibration of the bond being broken need be included, based on the premise that no other vibrational modes in the molecule are distorted by the reaction process. The symmetrical stretch for the ^{32}S—C bond was determined as 717 cm^{-1} from a Raman spectrum. The ^{34}S—C frequency was then calculated to be 711 cm^{-1}. From these frequencies and the Bigeleisen table of G-functions (p. 109) the TDF was found to be 1.003. Thus, for the dissociative process

$$\frac{k_\ell}{k_h} = (\mathrm{TIF})(\mathrm{TDF}) = 1.009 \times 1.003 = 1.012$$

In the formation of the associative transition state 3, the sulfur experiences an increased degree of bonding. We regard this process as equivalent to the formation of a diatomic molecule with the atoms on each side of the critical bond considered as one large group whose mass is the sum of the individual atomic weights. The TIF is computed as the square root of the ratio of the reduced masses of the heavy to the light isotopic fragments:

$$TIF = \left(\frac{\mu_h}{\mu_\ell}\right)^{1/2} = \left(\frac{\frac{111 \times 27}{138}}{\frac{109 \times 27}{138}}\right)^{1/2} = 1.009$$

The use of the diatomic model in which there is only one fundamental vibration reduces the TDF to $1 - G(u_j^{\neq})\ \Delta u_j^{\neq}$ [cf. Eq. (5.12)]. Since the bond involved is an S—C bond for which we assume the same frequencies as on p. 110, the TDF = 0.997. Thus, for an associative process

$$\frac{k_\ell}{k_h} = (TIF)(TDF) = 1.009 \times 0.997 = 1.005$$

The concerted process modeled on 2 would show an isotope effect intermediate between the associative and dissociative processes. The expected value is therefore

$$\frac{k_\ell}{k_h} = \frac{1.012 + 1.005}{2} = 1.0085$$

These calculations, while by no means exact, are quite useful in establishing the limits for the isotope effect of the three alternative mechanisms. This simplified treatment gives rise to a minimum value of k_ℓ/k_h for the dissociative model, since consideration of any additional frequencies would represent ground-state factors to be added to the $\Sigma_i G(u_i)\ \Delta u_i$ term, which is positive. The calculated k_ℓ/k_h arrived at analogously for the associative model, on the other hand, should be regarded as a prediction of its maximum value. In this case the additional frequencies to be considered would be transition-state factors to be added to the $\Sigma_i G(u_j^{\neq}) \Delta u_j^{\neq}$ term, which has a negative sign. Thus, such corrections to this simplified treatment would tend to increase slightly the k_ℓ/k_h differences between transition-state models $\underline{1}$ and $\underline{3}$. Summarizing, we find that the values of k_{32_S}/k_{34_S} at 198°C calculated for each of the models are

$\underline{1}$	$\underline{2}$	$\underline{3}$
1.012	1.008	1.005

The experimental ratio at 198°C is $k_{32_S}/k_{34_S} = 1.0040 \pm 0.0016$, confirming the associative transition state $\underline{3}$.

How does one measure these isotope effects with the accuracy required? High precision was realized through a relatively simple approach using a medium-resolution high-speed mass spectrometer linked with an appropriate data system. The question of accuracy, in the sense of absolute isotope ratio measurements, could present a problem because long-term instrument drift may make perfect agreement from day to day rare. For the α-methylallyl phenyl sulfide study the difficulty was obviated by always referencing the ratio to that found in the starting material.

KINETIC HYDROGEN ISOTOPE EFFECT

The kinetic isotope effect as it applies to hydrogen/deuterium has particular importance because hydrogen transfer reactions are very common. In general, isotope effects are much larger in the case of hydrogen where the ratios of isotopic masses are relatively enormous and give rise to much greater zero-point energy differences than are characteristic of heavy atom isotopes. Consider the three-center reaction

$$\text{A—H} + \text{B} \longrightarrow [\text{A---H---B}]^{\neq} \longrightarrow \text{A} + \text{B—H}$$

where A and B can be atoms or molecules. For example, the process depicted may be the transfer of hydrogen from a carbon-hydrogen bond in A—H to a carbon-hydrogen bond in B—H. Where the reaction proceeds via the transition state A---H---B, the major factor contributing to the generally lower reactivity of deuterium versus hydrogen substrates is commonly attributed to the difference in zero-point energy of corresponding bonds, namely, C—H versus C—D. Equation (5.12) also applies here as in the heavy atom isotope effect. Since we are concerned in hydrogen transfer reactions with breaking one bond while forming a new one in the transition state, we are interested in the influence of this factor on the relative magnitudes of terms A and B. The vibrational energies of H or D with respect to each of the atoms between which it is being transferred determine almost completely the magnitude of term B. In a first approximation the vibrational energies characteristic of all the other bonds in the configuration can be neglected with respect to the vibrational energy changes experienced by the transferring atom.

In Fig. 5.1 the potential energy is represented as a function of the extent of transfer of hydrogen from A to B. The same curve also applies to the transfer of deuterium from A to B. The main difference between H and D is the height of their vibrational levels. The force constant is governed by the shape of the potential well and, thus, will

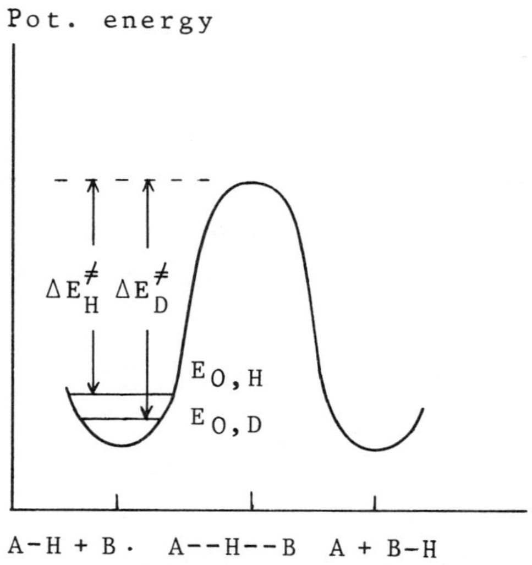

FIG. 5.1

not be different for hydrogen and deuterium. Since $\mu_D > \mu_H$, the vibrational levels are lower for D than for H. Most bonds, even at room temperature, occupy their lowest or zero-point level (n = 0). Quantitatively, a C—H stretching frequency of 2900 cm^{-1} correlates at ambient temperature with a zero-point energy of about 18 kJ/mol. The related C—D frequency is about 2100 cm^{-1}, and its zero-point energy 13 kJ/mol, 5 kJ/mol lower than for C—H bonds. Where there is no energy difference between the corresponding H and D transition states, the C—D bonds will require 5 kJ/mol higher activation to reach the activated complex. This energy difference corresponds to a rate ratio of

$$\frac{k_H}{k_D} = \exp \frac{5}{2.4} = 8.0$$

Assuming that the Arrhenius preexponential factors for corresponding H and D transfer reactions are identical, $A_H/A_D = 1$. The maximum deuterium isotope effects calculated in this way for a number of proton transfer reactions at ambient temperature are

R_3C—H	7.2	F—H	14.9
RO—H	10.2	Cl—H	7.2
RS—H	5.8	Br—H	5.9
		I—H	5.8

At higher than ambient temperature these maximum values of k_H/k_D decrease according to the Arrhenius-derived relationship

$E = [n + (1/2)]h\nu$

$n = 0, 1, 2, \ldots$

$\nu = (1/2\pi)\sqrt{f/\mu}$
f = force constant
μ = reduced mass

2900 cm^{-1} or $\nu = 87 \times 10^{12}\ s^{-1}$

$h\nu = 5.8 \times 10^{20}$ J/molecule or 35 kJ/mol

$(1/2)h\nu \simeq 18$ kJ/mol

kT at T = 300 K is equivalent to 4.14×10^{-21} J.

RT at ambient temperature = 2.4 kJ/mol

The preexponential factor A is sometimes called the frequency factor [Eq. (4.6)].

$$\frac{k_H}{k_D} = \frac{A_H}{A_D} \exp \frac{[\Delta E_a]_D^H}{RT}$$

where $[\Delta E_a]_D^H$ is the activation energy difference for corresponding H and D transfer processes. This term is equal to $[\Delta E_0]_D^H$, the zero-point energy difference for a linear H transfer. The maximum values can be expected when the exchange of D for H does not have an effect on the vibrational mode in the transition state, which is the symmetrical stretching vibration of the linear system A---H---B (cf. p. 80). This will be true when H is immobilized in this vibrational mode because the forces binding it to A and B, respectively, are balanced, i.e., when bond making and bond breaking energies are identical in the transition state.

$$\overleftarrow{A}\text{---H---}\overrightarrow{B}$$

The maximum isotope effect is not obtained when the degree of bonding in A---H and H---B is not identical, producing a transition state which, in this sense, is unsymmetrically structured. This can be associated with a reactantlike (A---H is stronger) or productlike (H---B is stronger) structure. Now, the stretch vibration will result in movement of the central atom and so become mass-dependent. Such an "unsymmetrical" symmetric vibration must have a zero-point energy different for H and D. This difference will give rise to a value of k_H/k_D less than the maximum. This can be perceived with the aid of the reaction profile diagram in Fig. 5.2.

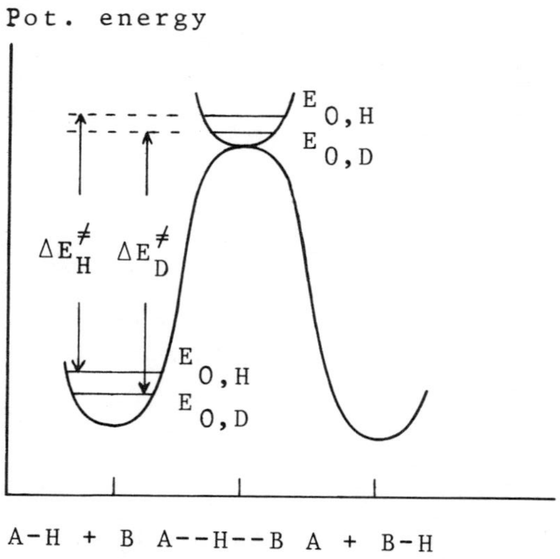

FIG. 5.2

Deformation vibrations can also reduce the isotope effect. In the transition state the restoring forces from H to A and to B are in about the same direction, and the frequency can even be larger than in the initial state, which results in a decrease of the ratio k_H/k_D. Because there are two bending vibrations (in perpendicular planes), the contribution will be double.

The isotope effect will also be smaller than its maximum value when the transition state for H transfer is bent. Then no balance of forces acting on H can be realized as in the linear case, and consequently the isotope effect will be smaller. The effect is also smaller because in this system the vibration undergoing change or removal in the activated complex is a bending instead of a stretching mode. Bending vibrations have much lower energies than stretching vibrations, and thus $[\Delta E_0]_D^H \rightarrow 0$. Therefore k_H/k_D approaches the value of A_H/A_D and temperature independence. Quantitative values have been calculated by O'Ferrall. For reasons that have not as yet been fully elaborated, under these circumstances A_H/A_D is > 1 and can become as great as 4.

A bent transition state is present in, e.g., the acetolysis of 3-methyl-2-butyl tosylate; k_H/k_D varies only slightly with temperature: 1.93 ± 0.10 at 75°C and 2.06 ± 0.08 at 25°C.

$$
\begin{array}{c}
\text{X} \quad \text{H} \\
| \quad\quad | \\
\text{H}_3\text{C}-\text{C}-\text{C}-\text{CH}_3 \\
| \quad\quad | \\
\text{CH}_3 \quad \text{OTs}
\end{array}
\;\longrightarrow\;
\left[
\begin{array}{c}
\text{X} \\
\text{H}_3\text{C} \diagdown \overset{+}{\diagdown} \\
\text{C}\text{---}\text{CHCH}_3 \\
\text{H}_3\text{C} \diagup
\end{array}
\;\;^{-}\text{OTs}
\right]^{\ddagger}
\;\longrightarrow\;
\text{products} \qquad \text{X = H,D}
$$

initial state

$$\text{A}--\text{H} \quad\downarrow\uparrow$$

transition state

$$\text{A} --- \text{H} --- \text{B} \quad\downarrow\uparrow$$

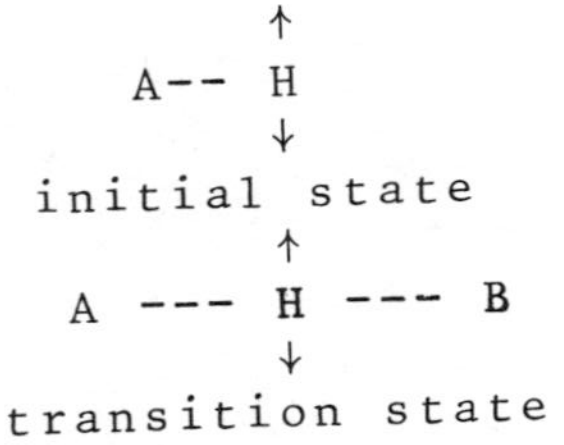

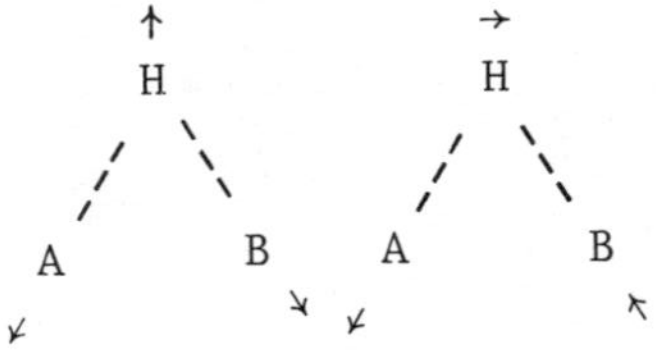

R. A. More O'Ferrall, J. Chem. Soc., B $\underline{1970}$:785.

S. Winstein and J. Takahashi, Tetrahedron $\underline{2}$:316 (1958).

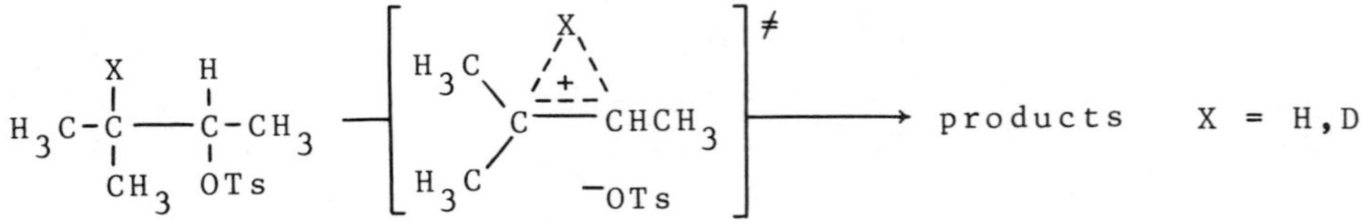

Recent calculations by C. F. Wilcox, I. Szele, and D. E. Sunko, Tetrahedron Lett. $\underline{1975}$:4457.

Another example is found in the thermolysis of an amine oxide:

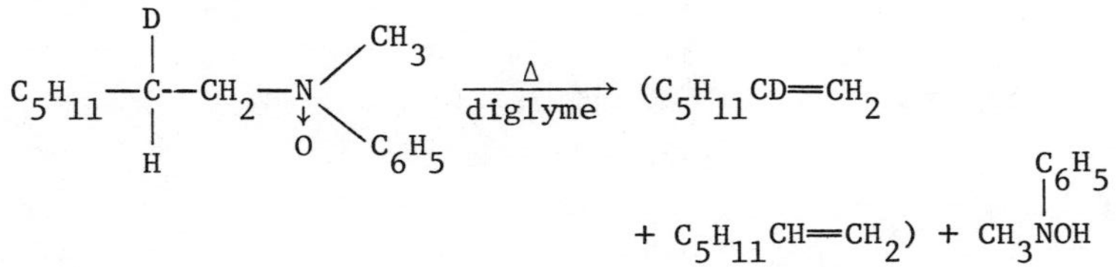

Over the appreciable temperature range of 363 to 483 K, k_H/k_D (= 2.209) did not change significantly, and thus $[E_a]_D^H$ = 0.00 kJ/mol. The ratio A_H/A_D = 2.209 ± 0.007. The conclusion to be drawn from these data is that the transition state of this well-known concerted reaction is bent.

TUNNEL CORRECTION

As mentioned on pages 90-91 hydrogen, because of its low atomic weight, deviates from classical behavior. There exists a finite probability for its surmounting an energy barrier even if it possesses less than the energy corresponding to the barrier height. The phenomenon is a direct consequence of the Heisenberg principle which states that $\Delta p\ \Delta x \geq h/2\pi$, where Δp is the uncertainty in the momentum and Δx the uncertainty in locating the particle. The deviation from classical behavior is larger for hydrogen than deuterium and can become an important factor determining the magnitude of k_H/k_D. For atoms of greater mass it is of relatively no importance.

$p = mv$

The uncertainty in the position of a particle is of the order of the De Broglie wavelength, $\lambda = h/mv$, where h is Planck's constant and m and v are the mass and velocity of the particle. For a proton (or hydrogen atom or hydride for that matter) moving with thermal velocity at ordinary temperatures, this wavelength is in the range of 10^{-8} to 10^{-9} cm. Since this is comparable to the width of the energy barrier in proton transfer reactions, the system must be treated quantum mechanically.

To place the tunnel effect on a quantitative basis the preferred treatment, given by Bell, assumes the barrier can be approximated by a symmetric, truncated parabola (Fig. 5.3). Any problems that might result because of the discontinuity at the base of the curve will be unimportant unless the main contribution to the reaction rate is made by particles of low energy, a condition that is not likely.

R. P. Bell, Discuss. Faraday Soc. 39:16 (1965); Proton in Chemistry (Cornell Univ. Press, Ithaca, N.Y., 1973), p. 270.

It is helpful to express the curvature at the top of the barrier by a frequency

$$\nu = \frac{E^{1/2}}{\pi a(2m)^{1/2}} \qquad (5.14)$$

where E is the barrier height, a is the half-width of the parabola, and m is the mass of the proton or deuteron. The frequency ν corresponds to that with which a particle of mass m would vibrate in a parabolic potential energy well having the same curvature as the barrier. Its imaginary counterpart iν corresponds to the antisymmetric mode (so-called reaction coordinate). It will be seen that the curvature of the barrier top also expresses a measure of the barrier thickness; i.e., low curvature corresponds to a wide barrier.

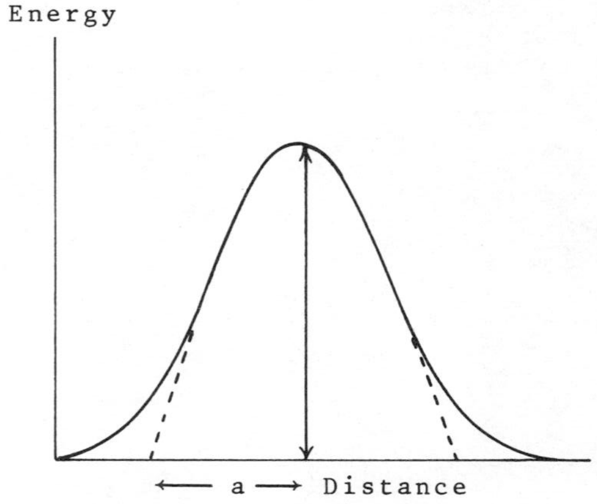

FIG. 5.3

The probability G of a particle of energy W traversing the barrier is given by the continuous function

$$G(W) = \left[1 + \exp \frac{2\pi(E - W)}{h\nu}\right]^{-1} \qquad (5.15)$$

where h is Planck's constant. The curve so defined is symmetrical about the point W = E, G(W) = 1/2 (Fig. 5.4). Averaging this probability over all possible thermal energies given by a Boltzmann distribution allows the formulation of the tunnel effect correction factor Q:

$$Q = \frac{k_{QM}}{k_{class}} = \exp \frac{E}{kT} \int_0^\infty \frac{1}{kT} \exp \frac{-W}{kT}\, G(W)\, dW \qquad (5.16)$$

using the QM and class subscripts to denote the quantum mechanical and classical parameters, respectively. Exact evaluation gives

$$Q = \frac{(1/2)u}{\sin(1/2)u} - u \exp\left(\frac{E}{kT}\right)\left(\frac{y}{2\pi - u} - \frac{y^2}{4\pi - u} + \frac{y^3}{6\pi - u} - \cdots\right) \qquad (5.17)$$

For chemical reactions at ordinary temperatures, y exp (E/kT) << 1, and the correction factor simplifies to

$$Q = \frac{(1/2)u}{\sin(1/2)u} \qquad (5.18)$$

At this juncture it seems useful to reiterate part of the reasoning discussed earlier in this analysis. Thus, as a direct result of the Heisenberg Uncertainty Principle, the incidence of tunneling is a function of the width relative to the height of the potential barrier.

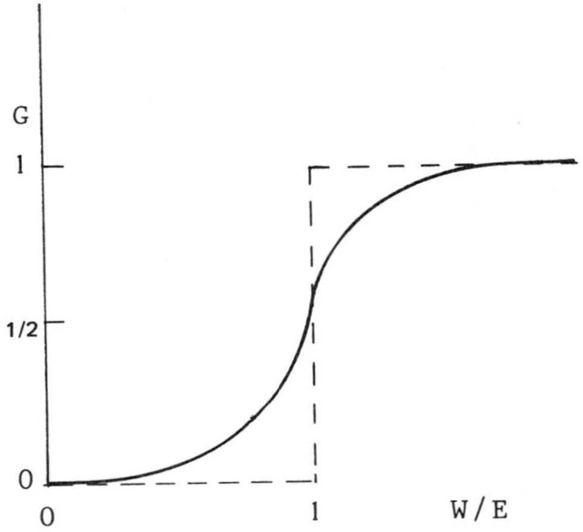

FIG. 5.4 Probability of traversing an energy barrier as a function of W/E for classical (broken line) and quantum (solid line) mechanics.

k = Boltzmann's constant

$u = \frac{h\nu}{kT}$

$y = \exp \frac{-2\pi E}{h\nu}$

$\Delta E\ \Delta t \geq \frac{h}{2\pi}$

When ΔE is larger, the lifetime of the particle in a certain state is smaller; i.e., the uncertainty in locating the particle is higher.

For a typical parabolic barrier shape (Fig. 5.5), as progress is made along the reaction coordinate the rising vibrational energy of the critical bond is attended by increasing uncertainty in the location of the hydrogen. Simultaneously, the width of the barrier is diminishing. For some pathways there exists a point on the reaction coordinate at which the width of the barrier is equal to the uncertainty in location of the increasingly energetic hydrogen undergoing transfer. Under such circumstances we can anticipate a significant degree of tunneling. In other words, a significant proportion of the hydrogens traveling this route toward bond rupture will be found to have penetrated to the product slope of the barrier without ever having experienced the activation energy required for scaling the classical top of the barrier.

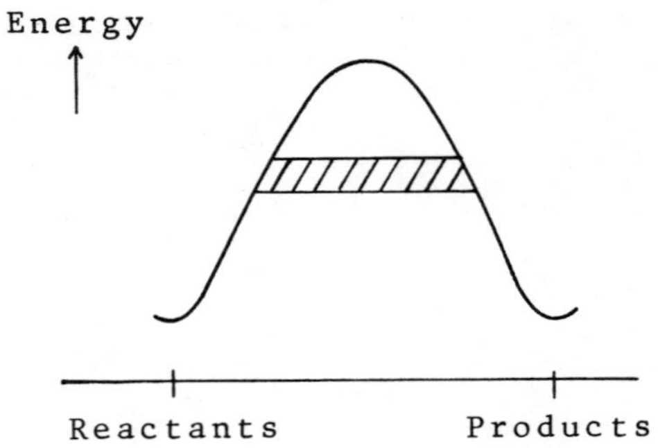

FIG. 5.5 Main tunneling region.

In the case of an unsymmetrical barrier, i.e., a reaction step with $\Delta H_0 \neq 0$, the degree of tunneling will be less than that encountered with a symmetrical barrier. The proton can only tunnel through that part of the barrier which is above both the initial and final states. It therefore becomes necessary to modify ν in Eq. (5.14) to take this point into consideration by defining

$$E = E_a - \Delta H_0 \tag{5.19}$$

where E now refers only to that part of the barrier which lies above the initial and final states (Fig. 5.6).

The shaded areas in Fig. 5.6 represent regions associated with probability for tunneling to occur. An unsymmetrical barrier is seen to possess a relatively smaller region.

In terms of Arrhenius parameters E_a and A [Eq. (4.6) and page 67] defined by the equations

$$E_a = RT^2 \frac{d \ln k}{dT} \tag{5.20}$$

$$A = k \exp \frac{E_a}{RT} \tag{5.21}$$

it follows that since $Q = k_{QM}/k_{class}$,

$$E_{a,QM} - E_{a,class} = RT^2 \frac{d \ln Q}{dT} \tag{5.22}$$

$$\frac{A_{QM}}{A_{class}} = Q \exp \frac{E_{a,QM} - E_{a,class}}{RT} \tag{5.23}$$

Substitution of an expression for Q such as the simplified one of (5.18) yields

$$E_{a,QM} - E_{a,class} = RT\left[\frac{1}{2} u(\cot \frac{1}{2} u) - 1\right] \tag{5.24}$$

$$\frac{A_{QM}}{A_{class}} = \frac{(1/2)u}{\sin(1/2)u} \exp\left[\frac{1}{2} u(\cot \frac{1}{2} u) - 1\right] \tag{5.25}$$

These modifications of the basic Arrhenius equation lead to two distinctive experimental criteria for the occurrence of tunneling:

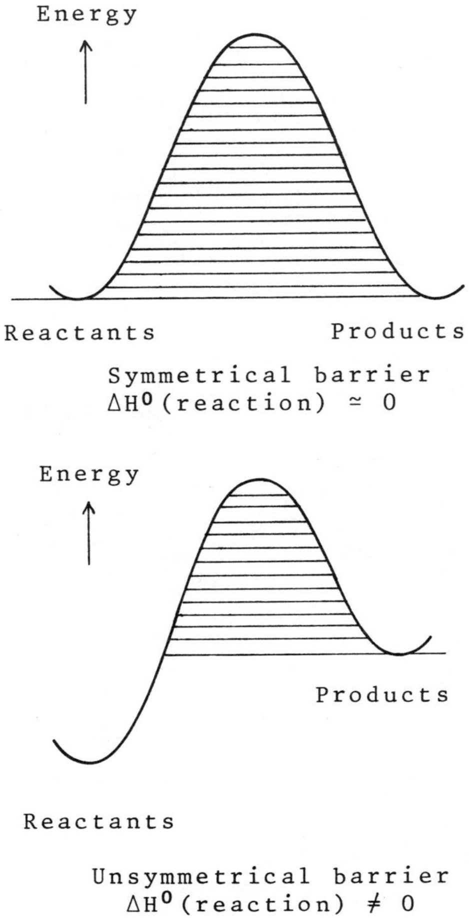

FIG. 5.6

1. Nonlinear Arrhenius plots. The plots in Fig. 5.7 show the occurrence of an upward trend from linearity with decreasing temperature. The reason for this increasing positive deviation is that, although the fraction of particles with energy greater than the barrier height decreases, there is still a finite chance for the particle to penetrate through the barrier depending on its width. The tunnel correction is most commonly observed at lower temperatures; cf. Eq. (5.18), where $u = h\pi/kT$.

2. Anomalous isotope effects

(a) The tunnel correction is larger for hydrogen than for the heavier deuterium. Thus k_H/k_D is larger than it would be in the absence of tunneling. A value of k_H/k_D appreciably exceeding the values given on page 114 could be indicative of tunneling.

(b) A difference in observed Arrhenius activation energies $E_a^D - E_a^H$ (see Fig. 5.7) greater than ≈ 5 kJ/mol (p. 114) indicates the possibility of tunneling; i.e., $[\Delta E_a]_D^H > [\Delta E_0]_D^H$, the zero-point energy difference.

(c) The preexponential factor in the Arrhenius equation is greater for D transfer than for H transfer, $A_H < A_D$; see the intercepts of the dashed lines in Fig. 5.7. Any experimental ratio A_H/A_D smaller than 0.75 constitutes evidence for tunneling.

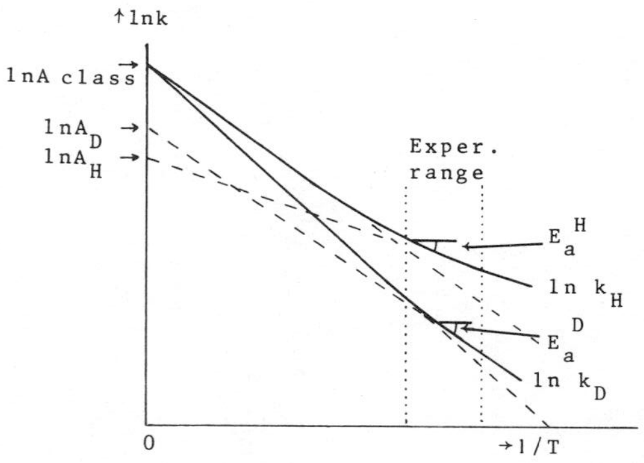

FIG. 5.7

Determination of the isotope effect over a temperature range permits the calculation of A_H/A_D and of $[\Delta E_a]_D^H$ in the equation

$$\frac{k_H}{k_D} = \frac{A_H}{A_D} \exp \frac{[\Delta E_a]_D^H}{RT} \qquad (5.26)$$

These quantities [just like $\Delta S^{\neq}$ and $\Delta H^{\neq}$ in Eq. (4.33) are indispensable to realizing a fundamental understanding of the mechanism of hydrogen transfer processes. Two examples are considered below.

For the Arrhenius equation, see (4.5) and (4.6).

The retroene thermolysis of each of the following substrates has very similar activation parameters despite large differences in the acidities of their O—H bonds and in the hybridization of the atoms in the pericyclic transition states. The isotope effects measured over an approximately 100° temperature range in the three cases are for all purposes identical; $A_H/A_D \simeq 1.0$ and $[\Delta E_a]_D^H \simeq 5.4$ kJ/mol, which is close to the computed maximum value for the O—H bond. These data require that the hydrogen is transferred linearly between oxygen and the variably

H. Kwart and M. C. Latimore, J. Amer. Chem. Soc. 93:3770 (1971).

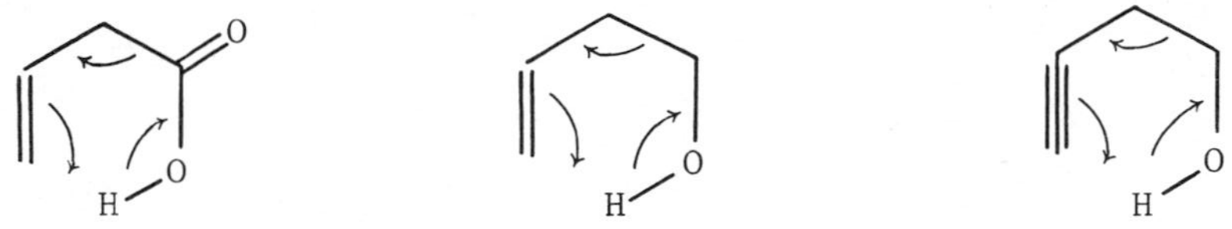

Vinylacetic acid 3-Butene-1-ol 3-Butyne-1-ol

unsaturated carbon centers separated by large distances in a broad activation barrier, as discussed on page 118.

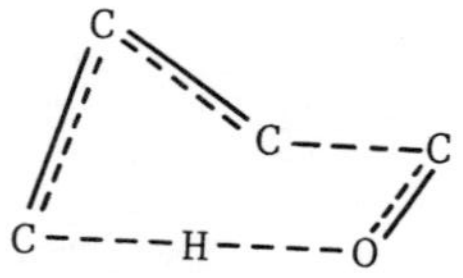

A similar situation exists in the thermolysis of normal aliphatic sulfoxides in bis(2-ethoxyethyl) ether as solvent. The isotope effect

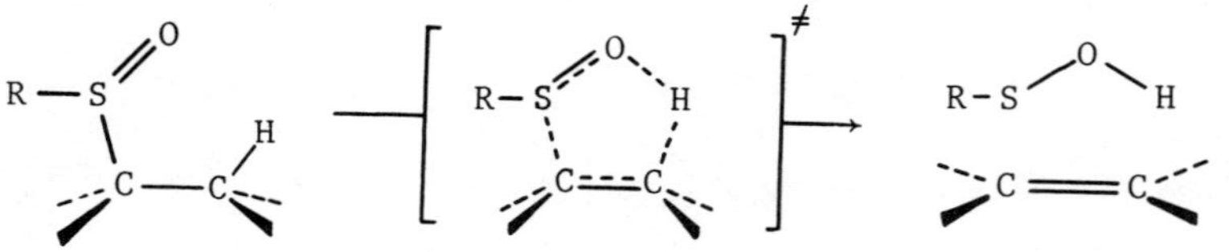

$R = C_5H_{11}$

H. Kwart, T. J. George, R. Louw, and W. Ultree, J. Amer. Chem. Soc. 100:3927 (1978).

was measured over the temperature region from 403 to 503 K. From Arrhenius plots $[\Delta E_a]_D^H$ was determined to be 4.8 kJ/mol, and A_H/A_D = 0.76. These data require H transfer in a planar pericyclic process. It will be remembered (p. 117) that the temperature-independent isotope effect ($[\Delta E_a]_D^H = 0$; $A_H/A_D = 2.209$) which characterized the analogous five-membered cyclic process of amine oxide thermolysis was held to indicate a bent transition state in sharp contrast to the above sulfoxide thermolysis.

The n-pentylethyl sulfoxide thermolysis also stands in contrast to the thermolysis of the highly branched analogs like tert-butylethyl sulfoxide. Measurements of the temperature dependence of the isotope effect in this decomposition reaction clearly show a large degree of tunneling in the course of hydrogen transfer accompanied by cleavage:

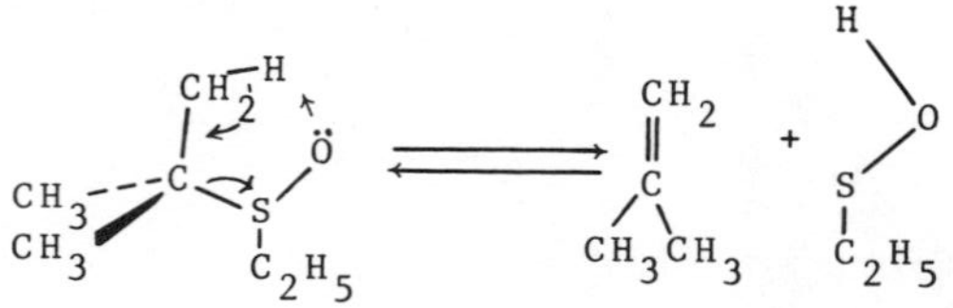

	E_a (kJ/mol)	A (s^{-1})
$C_4H_9SOC_2H_5$	123	1.7×10^{12}
$C_4D_9SOC_2H_5$	136	24×10^{12}
	$[\Delta E_a]_D^H = 13$ kJ/mol	$A_H/A_D = 0.07$

J. W. A. M. Janssen and H. Kwart, J. Org. Chem. 42:1530 (1977).

With the aid of models it can be seen that in the highly branched, hindered sulfoxide the oxygen atom is always close to one of the β(C—H) bonds. This situation results in a narrowing of the reaction barrier and, consequently, an increased probability of tunneling.

ACTIVATION BARRIER SHAPE AND DIMENSIONS

When the merger of the potential energy functions (p. 70) for the A---H and H---B bonds results in a large distance between the two minima, the barrier is broad, of low curvature at its top, and relatively high. See Fig. 5.8. The low curvature corresponds to a low imaginary frequency for the reaction coordinate mode (p. 118). Such a barrier engenders a near zero tunnel correction. The isotope effect for such a barrier is largely determined by zero-point energy differences. Simply put, the thicker the barrier, the less the probability of tunneling and the greater the extent of control of k_H/k_D exerted by zero-point energy factors. An example is given on p. 124.

When the merger of the potential energy functions for the A---H and H---B bonds results in a small distance separating the two minima, the

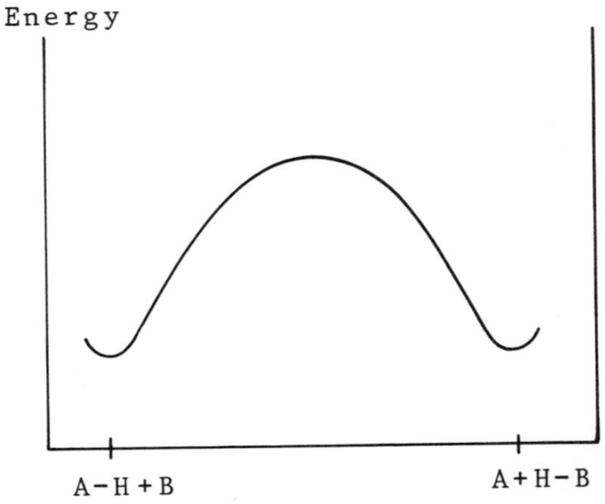

A-----B distance large

FIG. 5.8

barrier is thinner, characterized by a steeper curvature at the top, and of relatively lower height. See Fig. 5.9. For these circumstances a large tunnel correction can be anticipated. E_a^D will always be greater than E_a^H (Fig. 5.7), because the heavier mass atom is inclined to tunnel at higher energy levels and because it rises from a lower initial state of zero-point energy. An example of this situation has been presented on the preceding page. This is, the barrier to reaction E_a^H will be considerably lowered from the value it is possible to calculate (Bell-Caldin) for the same process in the absence of tunneling. A typical list of reactions in which tunneling is observed and the relative importance of this factor in lowering the reaction barrier is given in Table 5.1.

The assumption of a classical process of H transfer in competition with tunneling should not imply that the reaction can proceed by two entirely separate mechanisms under precisely the same conditions. This would be a violation of the rule that all molecules (particles) react according to the same quantum mechanical laws. All the molecules in a given reaction travel the same path. The tunnel correction is an expression of how much lower the energy level of reaction is driven compared to the hypothetical case where tunneling is not to be perceived.

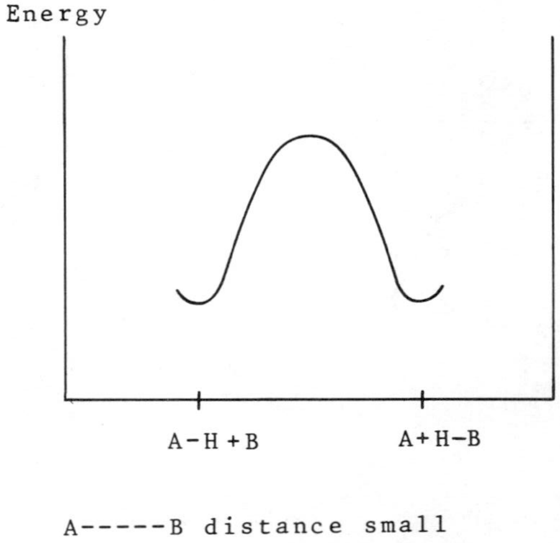

FIG. 5.9

TABLE 5.1 Relative Importance of Tunneling in Various Reactions (i.e., to what extent tunneling has lowered the barrier to reaction; energies are in kJ/mol)

	Experimental				Calculated by Bell-Caldin method			
Reaction, X = H or D	E_a^H	E_a^D	$[\Delta E_a]_D^H$	A_H/A_D	E_{class}^H	E_{class}^D	$[\Delta E_{class}]_D^H$	$[\Delta E_t]_D^{H*}$
Ionization of $4\text{-}NO_2—C_6H_4CX_2CN$ with EtO^-	38	46	8	0.1	43	47	4	4
Ionization of $(CH_3)_2CXNO_2$ with collidine	60	72	12	0.15	68	73	5	7
Enolization of $(CX_3)_2CO$ (in bromination) in H_2O	51	61	10	0.17	57	62	5	5
Enolization of acetone catalyzed by various bases (in bromination) in X_2O	49	54	5	0.44	55	56	1	4
$ClCH_2—CO_2^- + X_2O$	46	51	5	0.35	52	54	2	3
$F^- + X_2O$	61	71	10	0.04	75	75	0	10

*$[\Delta E_t]_D^H = [\Delta E_a]_D^H - [\Delta E_{class}]_D^H$ is the estimated difference in activation energy for classical and tunneling processes at a given temperature t.

SECONDARY DEUTERIUM ISOTOPE EFFECT

As outlined in the foregoing discussion, the origin of the kinetic isotope effect is found in changes in the quantized molecular vibrational levels. Nothing specifically requires a bond to the isotopic atom to undergo rupture in the course of reaction. A secondary isotope effect is said to occur as a consequence of bond breaking or making in which the isotopic atom is not directly involved, i.e., in which a bond to the isotopic atom is neither formed nor broken in the activated complex. An effect of significant magnitude can be perceived only for those reactions in which the vibrational patterns of the reactant and transition state do not match perfectly in a part of the molecule remote from the seat of reaction. This effect, just as the primary isotope effect, since it is dependent on the mass of the species involved, will be largest and most easily measured in cases involving the isotopes of hydrogen because of their relatively large mass difference.

The theory of kinetic secondary isotope effects is exactly analogous to that developed for isotope effects in general. Again, according to Bigeleisen, any isotope effect can be expressed as a function of vibrational frequencies alone. Thus, neglecting anharmonicity and vibrational-rotational interactions, setting the ratio of the transmission coefficients equal to unity and assuming all frequencies are high relative to kT/h, one obtains

$$\frac{k_H}{k_D} = \left(\frac{m_D^*}{m_H^*}\right)^{1/2} \exp\left\{\frac{-h}{2kT}\left[\sum (\nu_H^{\neq} - \nu_D^{\neq}) - \sum (\nu_H - \nu_D)\right]\right\} \qquad (5.27)$$

Equation (5.27) follows from Eq. (5.12) for very large values of u_i.

Since in secondary effects the isotopic label is not involved in motion along the reaction coordinate, the temperature-independent term drops out, giving [cf. Eq. (5.11)]

$$\frac{k_H}{k_D} = \exp\left\{\frac{h}{2kT}\left[\sum (\nu_H - \nu_D) - \sum (\nu_H^{\neq} - \nu_D^{\neq})\right]\right\} \qquad (5.28)$$

where the summation is taken over those vibrational frequencies which have been altered in an important way by deuterium substitution. In the case where degeneracies occur, each is counted separately.

For those systems in which a vibrational analysis is available for only the parent molecule and not the deuterated analog, a method has been devised for calculating the relative rate ratio by estimating the frequencies of the deuterated compound. In addition to those approximations implicit in Eq. (5.28) it is further assumed that the ratio of C—H frequencies to C—D frequencies is generally given by $\nu_H/\nu_D = 1.35$. This ratio is slightly lower than what would be predicted for hydrogen and deuterium bound to an infinitely heavy atom but is more appropriate for real systems. Including this assumption in Eq. (5.28) leads to the simple expression

$$\frac{k_H}{k_D} = \exp\left\{\frac{0.187}{T}\left[\sum_i (\bar{\nu}_H - \bar{\nu}_H^{\neq})\right]\right\} \qquad (5.29)$$

$\bar{\nu} = c\nu$

c = speed of light

$\bar{\nu}$ in cm^{-1}

where the summation is now carried out only over the vibrational frequencies of the referenced hydrogen in the reactant and the transition state. These equations are, therefore, essential to an interpretation of the rate of reaction of a molecule possessing at least one C—H bond, at a site removed from the center of reaction, relative to its deuterated analog in terms of a change in zero-point energy difference on going from the reactant to the transition state.

The magnitude of this zero-point energy difference between C—H and C—D bonds in their lowest vibrational levels in both the reactant and the transition state is a function of the respective force constants f for the various bonds. Again, the force constant is related to the frequency of a simple harmonic oscillator and the reduced mass of the bonded atoms.

Thus, the tendency for a reaction to proceed with the development of smaller force constants in the transition state than in the reactants would result in a decrease in the zero-point energy difference and consequently a value of $k_H/k_D > 1.0$, a so-called normal isotope effect. Conversely, a tendency toward larger force constants and, therefore, a greater zero-point energy difference in the transition state would be correlated with a relative rate ratio $k_H/k_D < 1.0$, sometimes called the inverse isotope effect.

In general, an isotope effect will be found whenever in the course of transforming reactant to transition state there is a change in the vibrational force constants of bonds to isotopic nuclei. Since the force constant changes can originate in several ways, the following classification scheme for secondary isotope effects has been adopted:

1. The α-effect or "secondary isotope effect of the first kind" occurs when a bond to an isotopic atom has undergone spatial reorientation.
2. The β-effect and other remote effects or "secondary isotope effects of the second kind" are those observed when the configuration about a bond to an isotopic atom is retained.

By way of example of the α-secondary isotope effect, Streitwieser et al. have shown that the rate of acetolysis of 1-d-cyclopentyl tosylate is slower than the undeuterated analog ($k_H/k_D = 1.15$). Assuming an sp^2 carbon developing in the transition state, they chose the aldehydic C—H vibrational frequencies as a model and calculated the corresponding C—D frequencies. The observed effect was attributed to the change of a tetrahedral C—H bending vibration to an "out-of-plane" deformation in the transition state. Since the latter vibration is weaker (less impeded and therefore a smaller force constant), a normal isotope effect is expected.

A. Streitweiser, J. Amer. Chem. Soc. 80:2326 (1958).

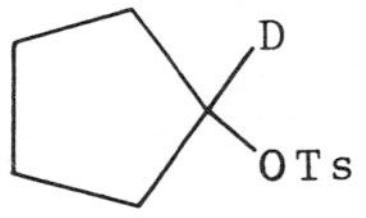

On the other hand, Denney and Tunkel have demonstrated that the addition of a variety of reagents to the double bond of trans-stilbene-d_2 and the undeuterated parent results in a relative rate ratio of less than unity ($k_H/k_D = 0.88$). They related this inverse effect to the increase in frequency of the same out-of-plane C—H bending mode occurring as the molecule leaves the planar ground state.

D. B. Denney and N. Tunkel, Chem. and Ind. 1959:1383.

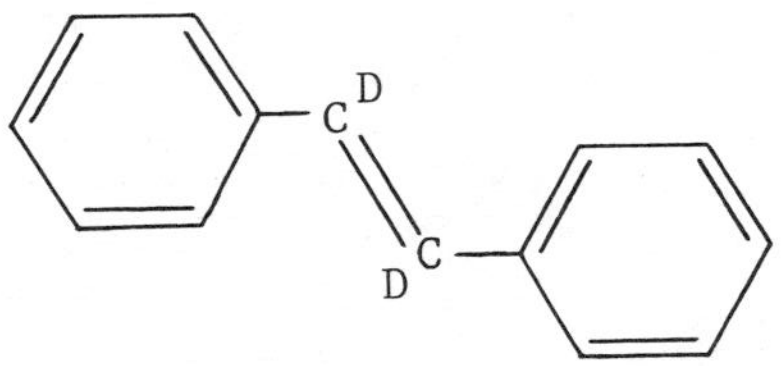

In general, then, the following can be said of α-secondary isotope effects: A rehybridization of the type sp^3 to sp^2, as in Streitwieser's example, will be accompanied by a decrease in force constants of α-C—H bonds resulting in a decrease in the zero-point energy difference and

$k_H/k_D > 1$; the sp^2 to sp^3 rehybridization, as in the stilbene example, will be attended by increased force constants and an inverse isotope effect.

β-Secondary isotope effects are associated with hyperconjugation and induction. In the solvolysis of tert-butyl chloride, Robertson et al. observed a normal isotope effect, $k_H/k_D = 2.387$ (1.10 per D). Evans and Lo performed a vibrational analysis of tert-butyl chloride and the corresponding cation and calculated an equilibrium isotope effect of $K_H/K_D = 2.358$. The excellent agreement suggests that the tert-butyl cation might serve as a model for the transition state in the solvolysis. The normal direction of this isotope effect is attributed to the hyperconjugative overlap of the C—H bond orbital with the electron-deficient p-orbital of the developing cation causing a weakening of the bond and lowering of the bending vibrational frequency of the bond.

L. Hakka, A. Queen, and R. E. Robertson, J. Amer. Chem. Soc. 87:161 (1965).

J. C. Evans and G. Y. Lo, J. Amer. Chem. Soc. 88:2118 (1966).

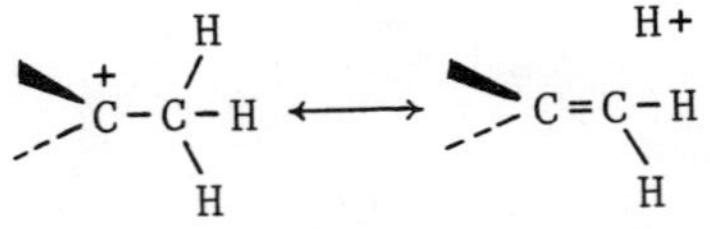

Induction is more difficult to document because it is directly observable only when hyperconjugative overlap has been structurally prevented. Shiner et al. have demonstrated that the solvolysis of the bicyclic tertiary chloride 2 is characterized by an inverse isotope effect. Since here hyperconjugation is structurally forbidden, an inductive effect on the developing cation in the transition state could be held responsible. The inverse isotope effect is clearly in line with the greater electron withdrawal of H versus D.

Inductive effect in CD_3COOH vs. CH_3COOH; E. A. Halevi, M. Nussin, and A. Zon, J. Chem. Soc. 1963:866.

V. J. Shiner and J. S. Humphrey, J. Amer. Chem. Soc. 85:2416 (1963).

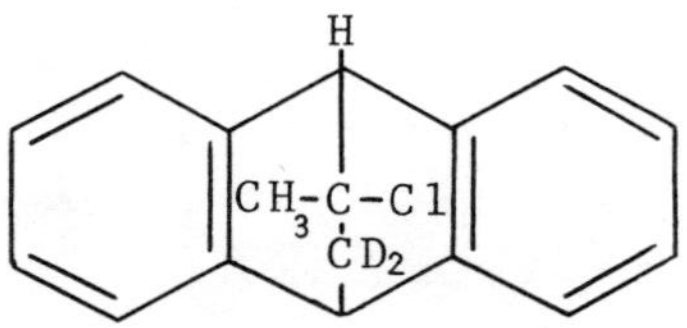

1

$k_H/k_D = 1.14$

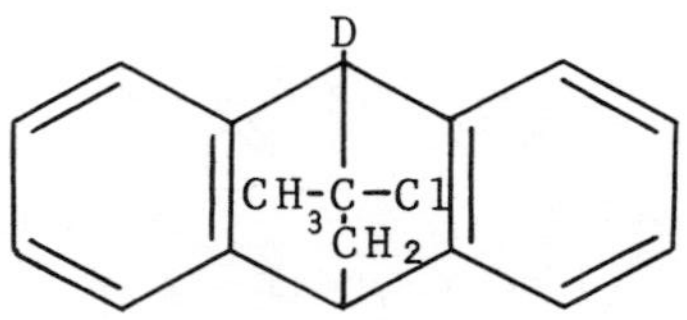

2

$k_H/k_D = 0.986$

As suggested by the discussion and shown in the examples, the size of secondary deuterium isotope effects is not very large. Whereas at 25°C a classical primary deuterium isotope effect as large as 18 can be expected, the maximum secondary effect on k_H/k_D is to increase or decrease its magnitude by 0.1 to 0.15 per deuterium. Because of the minute size of this effect, special care must be taken in both the experimental and theoretical treatments of it. Experimentally, for example, the simultaneous occurrence of a primary process can completely obscure the secondary one. The intervention of a polar solvent can so alter the transition-state structure as to make a correlation between experimental observations and a simple mechanism impossible. Temperature variation during or between experimental determinations renders the comparison of supposed similar effects meaningless [see also Eq. (5.29)]. In short, nothing but the most exacting experimental work and the most precise analytical determination of the isotope effect can provide useful data.

Similarly, from a theoretical standpoint, the assumptions made in the derivation of Eqs. (5.28) and (5.29) or the choice of an inappropriate transition-state model may impair the value of the kinetic data. The neglect of anharmonicity and rotation-vibration interaction as well as the assumption that electronic wave functions are essentially the same for isotopically related molecules and that their differences can be adequately described in terms of simple vibrational energy differences alone [all implicit in Eq. (5.28)] might be a source of significant error with so small an effect. The situation is further complicated because of the added approximation that all the vibrations are independent. Equation (5.29) therefore excludes contributions from coupled vibrations. Since an a priori evaluation of these assumptions is impossible, the general validity of the equations can only be determined by consideration of specific examples.

More importantly, the accuracy of the theoretically calculated k_H/k_D is a function of the proper assignment of vibrational frequencies to represent the transition-state structure chosen for evaluation. Once the critical frequencies have been identified, it becomes necessary to know how they change in going from reactant to transition state. In this manner we approach such basic questions as the following: How close is the transition state to the reactant/product? To what degree has planarity been lost/gained? To what extent has charge development progressed? What effect does the presence of formal charge have on the isotope effect?

Another application of the secondary deuterium isotope effect to the elucidation of transition state features is illustrated by the reaction of diphenylketene and styrene studied by Baldwin. The product of this [2 + 2] cycloaddition is formed in highly stereoselective fashion: cis-β-deuteriostyrene leads to 2,2,3-triphenyl-4-deuteriocyclobutanone having cyclobutyl hydrogens in a cis relationship. Secondary deuterium isotope effects for the reaction are normal for the α-position and inverse for the β-position:

J. E. Baldwin and J. A. Kapecki, J. Amer. Chem. Soc. 92:4874 (1970).

$C_6H_5-CD=CH_2$ $\quad k_H/k_D = 1.235$

$C_6H_5-CH=CHD$ $\quad k_H/k_D = 0.908$

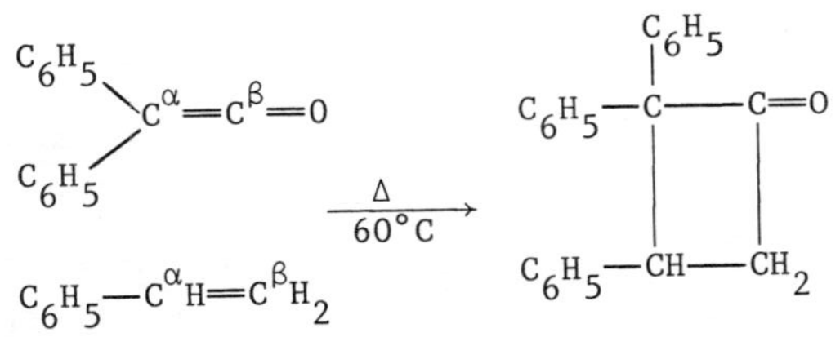

In undertaking an interpretation of these data, two earlier observations must be borne in mind:

1. In radical additions at olefinic centers the carbon becoming the new radical center shows no secondary deuterium isotope effect ($k_H/k_D \simeq 1.0$), retaining its nominal sp^2 hybridization in the activated complex.
2. In concerted Diels-Alder cycloadditions, where dienophiles are isotopically substituted at either end of the olefinic system, k_H/k_D-values very close to 1.0 support a transition state in which bond making seems to occur simultaneously at both reaction centers.

W. A. Pryor, R. W. Henderson, R. A. Patsiga, and N. Carroll, J. Amer. Chem. Soc. 88:1199 (1966).

Thus, the β-value ($k_H/k_D \approx 0.91$) is in accord with a transition state in which the hybridization at the β-position has progressed substantially toward full sp^3 (page 000). The effect of the α-position ($k_H/k_D \approx 1.23$) does not agree with the concerted model of the transition state, as found for the Diels-Alder, wherein both olefinic centers form bonds simultaneously ($k_H/k_D \approx 1.0$ or slightly less than 1.0). Nor does this result agree with prediction based on a diradical, two-step model.

Although an unequivocal explanation of the α-isotope effect cannot be ventured, one can infer that the p-orbital on the α-diphenylketene carbon is not disposed to overlap the p-orbital developing at the α-styrene carbon. However, the propinquity of the α-diphenylketene p-orbital to the α-styrene-hydrogen has considerably weakened the α-C—H bond. This may be regarded as a variation of a hyperconjugation effect in that it arises from similar electronic origins shown in the accompanying figure.

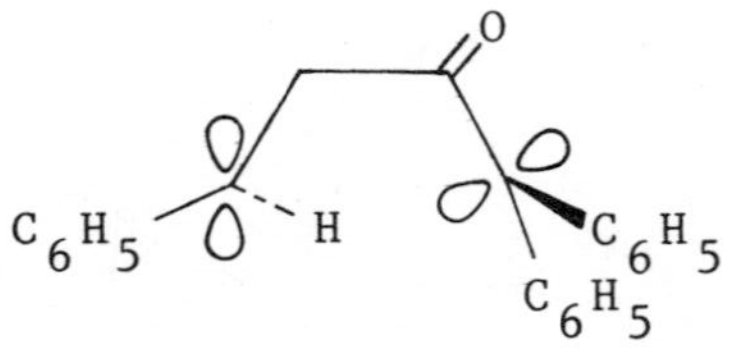

Similar results have been observed in the [2 + 2] cycloaddition of dimethyl azodicarboxylate with RO—CH═CH_2, where ═CH_2 versus ═CHD has k_H/k_D = 0.83 and CH═ versus CD═ has k_H/k_D = 1.12.

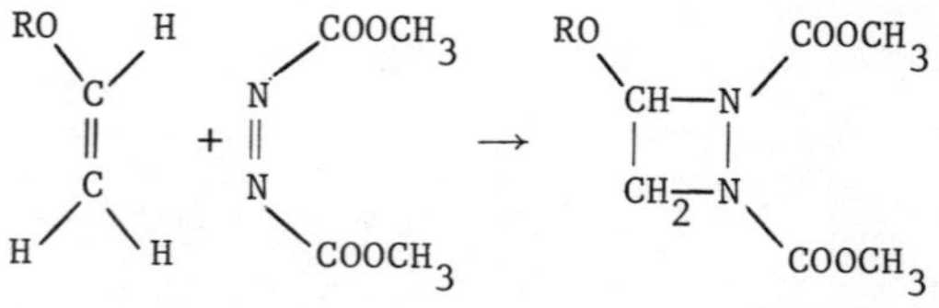

Through the isotope effect we are able to perceive events occurring at both the carbon seats of reaction in the transition state of so-called concerted [2 + 2] cycloaddition. Both of these carbons are establishing bonds to atoms of the other reaction component. In the cases considered above they achieve this bonding via dissimilar mechanisms.

6 Fast Reactions

Expansion of the range of kinetic measurement

Chapter 3 was concerned with various methods for measuring reaction rate. For instance, samples can be analyzed by titration, or the value of a physical parameter can be followed during the reaction. When the reaction takes a week, a day, or an hour to complete, samples can easily be withdrawn or readings made. However, sometimes the reaction proceeds so rapidly that accurate pursuit is barely possible. For classical measurements a half-life of the order of 10 s is the lower limit.

MODIFICATION OF CLASSICAL METHODS

For a reaction which is too rapid to follow in the classical way, there are several alternatives to be tried such as lowering the reaction rate by decreasing the temperature. However, low temperatures can give rise to technical problems, a case in point being that several common solvents such as benzene, water, and acetic acid solidify near 0 deg. In general, the reaction rate is given by an equation such as

$$v = kc_1c_2$$

The rate decreases when either k decreases by lowering the temperature or when the concentrations are reduced. An example of the latter procedure is found in the bromination reaction of aniline mentioned on page 57:

$$v = kc_{aniline}c_{Br_2}$$

For concentrations of 0.1 mol/liter of aniline and 0.1 mol/liter of bromine the rate is too fast and cannot be measured directly. But the rate can be reduced to a reasonable value by decreasing the concentration of aniline or bromine or both. When the reaction takes place in the presence of an acid, the following equilibrium occurs:

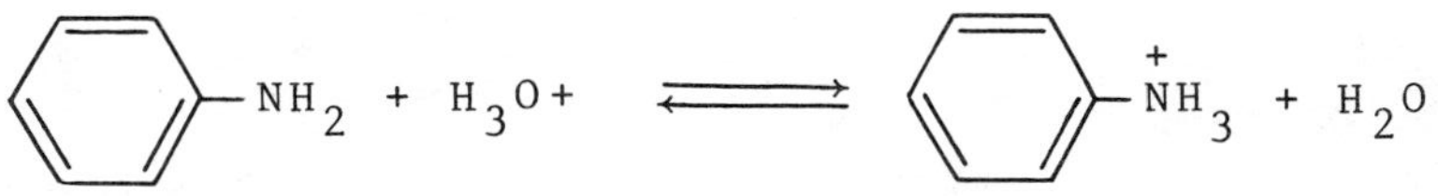

Only the free aniline will be brominated. The acid concentration regulates the concentration of free aniline and, consequently, the reaction rate. In the same way, the concentration of free bromine depends on the concentration of bromide ions.

A small concentration of bromine can also be obtained by electrolysis. When electrolyzing an aqueous solution of a bromide salt, the quantity of electrical energy passed is equivalent to the liberated bromine. The bromine concentration is increased by electrolysis and decreased by the reaction with aniline. Soon a stationary state is reached which is observable by polarography. The reading of the polarograph is a measure of the bromine concentration. This method was applied in the bromination of tyrosine. See Fig. 6.1. The tyrosine was present in large excess:

$$v = kc_{tyr}c_{Br_2}$$

In the stationary state the rate v is equal to the rate of bromine consumption, which in turn is equal to the rate of bromine generation by the electric current. As c_{tyr} is known and c_{Br_2} is measured polarographically, the rate constant k can be calculated.

$$Br_2 + Br^- \rightleftharpoons Br_3^-$$

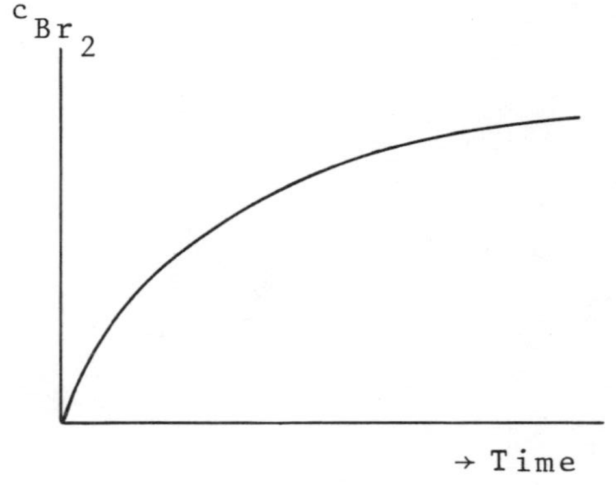

FIG. 6.1

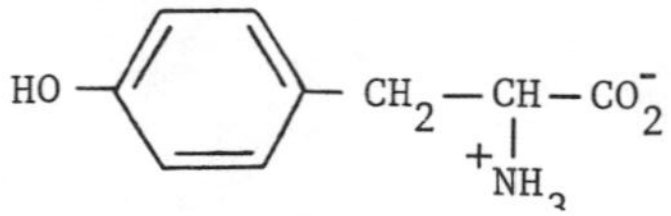

G. O'Dom and Q. Fernando, Anal. Chem. 37:893 (1965).

FLOW METHODS

Sometimes a decrease of temperature or concentration is impossible or undesirable. Special techniques have been devised to extend rate measurements to shorter half-lives. The first of these techniques is the

For details, see B. Chance, in Investigation of Rates and Mechanisms of Reactions, Part II, G. G. Hammes, ed. (Wiley-Interscience, New York, 1974), Chap. 2.

flow method, introduced in 1923 by Hartridge and Roughton.

Just as with many other new techniques, the reason for its introduction was a biochemical problem, namely, the rate of reaction between aqueous solutions of oxygen and hemoglobin. This reaction, taking place in vivo, is completed in a few seconds. The flow technique enables rate measurements down to 0.001 s half-life time. For a sketch of the apparatus, see Fig. 6.2. For the reaction between A and B both compounds are dissolved and the solutions stored in separate cylinders. Mixing is effected by opening the stopcock and moving the plungers downward. While A and B react, the mixture runs through the reaction tube. Progress of the reaction is usually followed by measuring the absorption in the ultraviolet. The reaction tube is placed in the beam of the ultraviolet spectrometer, and the extent of absorption at positions (1), (2), (3), etc., is estimated. Since the flow rate is known, the known distance between the mixing chamber and the measuring point provides the reaction time axis. The rate constant can be calculated in the usual way from the relation between absorption and time. Instead of the ultraviolet spectrum, another physical parameter may be used. For example, at various stations along the reaction tube pairs of electrodes can be assembled to measure the conductivity. More simply, a single pair of electrodes may be used and the flow rate varied.

The latter method has been applied for calculating the rate of decomposition of H_2CO_3 into H_2O and CO_2. The cylinders are filled with aqueous solutions of $NaHCO_3$ and HCl, both 0.005 mol/liter. Upon mixing, H_2CO_3 and Na^+ and Cl^- ions are formed almost instantaneously.

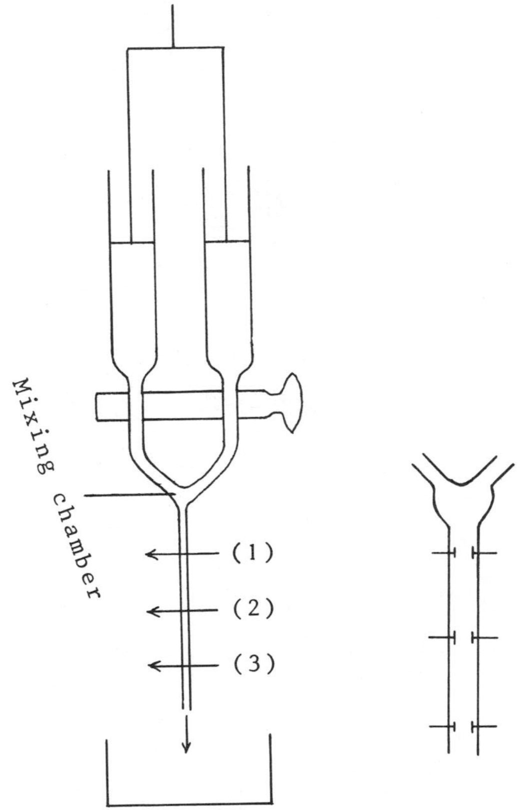

FIG. 6.2

R. G. Pearson, R. E. Meeker, and F. Basolo, J. Amer. Chem. Soc. 78:709 (1956).

The reduction in conductivity can be readily correlated with H_2CO_3 decomposition. The rate of change of conductivity therefore permits direct calculation of the rate constant desired. At 25°C, $k = 27.9\ s^{-1}$ and $t_{1/2} = 0.025$ s.

$$v = kc_{H_2CO_3}$$

A glass electrode can also be mounted in the tube. The pH of the H_2CO_3 solution can be measured, or if an excess of $NaHCO_3$ is used, the pH of the buffer $NaHCO_3$–H_2CO_3. Thus, the pK of H_2CO_3 can be calculated under conditions where the flow rate is fast enough and the H_2CO_3 is decomposed only to a minor extent (<< 0.02 s). The pK appears to be 3.58. This signifies that H_2CO_3 is a much stronger acid than is suggested by the apparent pK estimated for a solution of CO_2.

J. Meier and G. Schwarzenbach, Helv. Chim. Acta 40:907 (1957).

$$CO_2 + 2H_2O \rightleftharpoons HCO_3^- + H_3O^+$$

Apparent pK = 6.46

A similar example is the determination of the number of anionic carboxylate groups on the surface of coiled protein molecules dissolved in water. The normal procedure for determining carboxylate ions is by acid titration. Acid, however, denatures the protein, causing uncoiling and exposure of a much larger number of carboxylate ions. Uncoiling may require only 1 min, yet normal titration techniques cannot meet this pace. Application of the flow technique affords a solution to this problem.

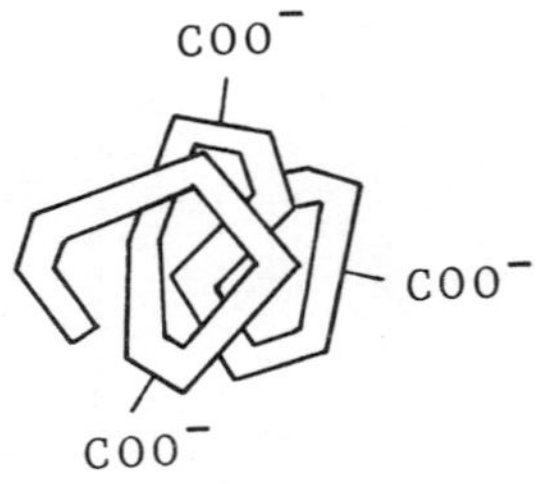

Variations of this flow technique have been developed. For instance, the flow can be stopped suddenly and the conductivity or another parameter displayed on an oscilloscope as a function of time. An advantage of this stopped flow compared with the continuous flow technique is that the volumes of the solutions required are much smaller. An example is the acid catalyzed addition of water to ethylthioethyne:

$$C_2H_5—S—C{\equiv}CH + H_2O \xrightarrow{H^+} C_2H_5—S—\overset{\overset{\displaystyle O}{\|}}{C}—CH_3$$

For pH 3 or 4, the rate of reaction can be determined in the classical way by measuring the increase of ethyl thiolacetate absorption at 233 nm. However, in 0.5 mol/liter of $HClO_4$ at 25°C the half-life is less than 0.15 s. In the acidity range 0.5 to 3.0 mol/liter of $HClO_4$, the solutions of ethylthioethyne and perchloric acid are mixed in a stopped flow apparatus, and the ultraviolet absorption, simultaneously displayed on an oscilloscope with a long retention trace, can be determined by a photographic procedure. In 3 mol/liter of $HClO_4$ at 25°C the half-life of the substrate is 0.0012 s.

W. F. Verhelst and W. Drenth, J. Org. Chem. 40:130 (1975).

FLASH PHOTOLYSIS

Flash photolysis can be used for studying fast reactions of particles which are generated photochemically in the gas phase or in the liquid phase. By a sudden discharge of a series of condensers, a short but very intense light pulse is generated. The flash time is approximately 10^{-5} s and with a laser even shorter, ca. 10^{-8} s. In this way a considerable number of particles in the excited state or of a photochemical product can sometimes be formed. These particles are analyzed by a second but much less intense beam which is passing through the reaction

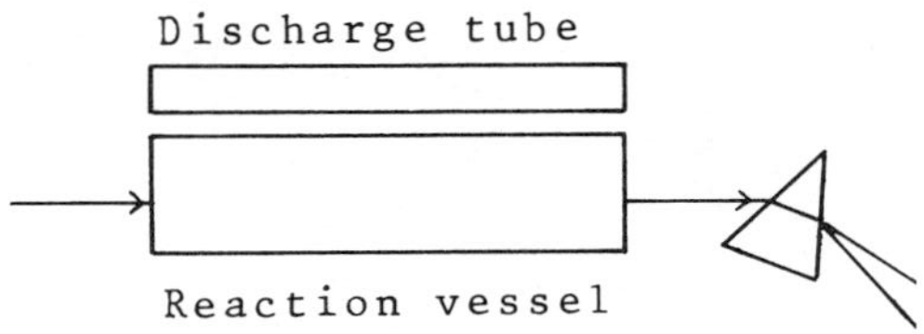

1 ns = 1 nanosecond = 10^{-9} s

Flash photolysis was developed by George Porter and Ronald G. W. Norrish (Nobel Prize, 1967).

vessel at right angles to the direction of the activating beam. Analysis can be carried out at various times after the intense flash, and the rate of disappearance of the photochemically generated particles can be estimated in this manner.

For example, in the photochemical decomposition of diazomethane, singlet carbene is formed initially. This singlet transforms to triplet carbene very rapidly. The energy released in this reaction is transferred during collision with an inert gas molecule, with the result that more rapid transformation of singlet → triplet is realized at higher inert gas pressure. By measuring the triplet concentration 6.5 μs after the initial pulse and with various helium concentrations, the second-order rate constant can be calculated:

$$k_m = (3.0 \pm 0.7) \times 10^{-13}\ cm^3\ molecule^{-1}\ s^{-1}$$

G. Porter and M. A. West, in Investigation of Rates and Mechanisms of Reactions, Part II, G. G. Hammes, ed. Wiley-Interscience, New York, 1974), Chap. 10.

$$CH_2N_2 \xrightarrow{h\nu} CH_2 + N_2$$

$$^1CH_2 + M \xrightarrow{k_m} {}^3CH_2 + M$$

W. Braun, A. M. Bass, and M. Pilling, J. Chem. Phys. 52:5131 (1970).

ION CYCLOTRON RESONANCE

Normally, in a mass spectrometer the gas pressure is so low and the path length so short that collisions between particles can be neglected. By increasing the gas pressure and the path length, reactions between ions and neutral molecules appear. The path length is increased by applying the principle of the cyclotron. The path of the ions resembles a tightly coiled helix of many turns wound in a plane perpendicular to the magnetic field. The frequency of the circular movement depends only on the ratio charge/mass of the ion and on the strength of the magnetic field. Between electrodes stationed at both sides of the

$$A^+ + B \longrightarrow C^+ + D$$

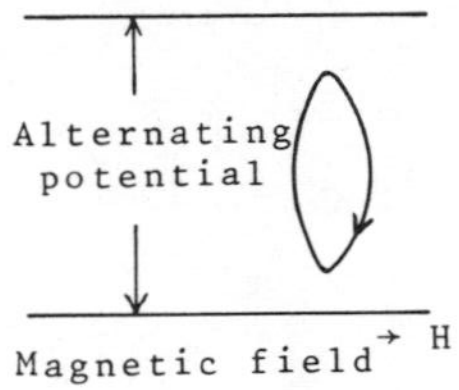

See the review by C. J. Drewery, G. C. Goode, and K. R. Jennings, M.T.P. Int. Rev. Sci. Phys. Chem. 5:183 (1972).

circular ion pathway an alternating electrical potential is applied. When the frequency of the applied potential is equal to the frequency of the circular movement, the ions absorb energy from the alternating field. This absorption can be detected by suitable instrumentation. By measuring the energy of absorption at constant magnetic field and at each different frequency of the alternating electric field, the signals of the different ions appear. Consider, for example, A^+ and C^+. The intensity of C^+ can be measured as a function of the pressure of B from which the kinetic behavior of the reaction can be inferred.

$$A^+ + B \longrightarrow C^+ + D$$

It is possible to apply two frequencies simultaneously. If with one frequency C^+ is measured, then its intensity might increase when energy is supplied to A^+ via the other frequency. Through this supply of energy the reactivity of A^+ is enhanced, and the rate of formation of C^+ increases proportionally. From such behavior it can be concluded that C^+ is formed from A^+.

Analogous reactions can be carried out for both negative and positive ions. An example involving negative ions is the following proton transfer:

$$ROH + R'O^- \rightleftarrows RO^- + R'OH$$

If this reaction proceeds to the right, ROH is a stronger acid than R'OH. Brauman and Blair observed the following order of acid strengths in the gas phase:

J. I. Brauman and L. K. Blair, J. Amer. Chem. Soc. 92:5986 (1970).

R = tert-butyl > isopropyl > ethyl > methyl > water

This sequence is the opposite of what has been observed for these alcohols in solution. The acidity order in solution has often been attributed to the inductive electron-releasing effect of the alkyl group. In the gas phase, however, the higher stability of the larger ions is most probably due to their higher polarizability. In the liquid phase, solvation is the predominant influence. Larger ions are solvated much less effectively.

RELAXATION METHODS

An equilibrium such as

$$A + B \rightleftarrows C + D$$

has a rate to the right, v_r, and a rate to the left, v_ℓ. If the equilibrium is established slowly, classical methods can be used for the measurement of these rates. If the equilibrium is established rapidly, sometimes relaxation methods allow measurement of the rates v_r and v_ℓ as follows.

Consider the equilibrium

$$HA + H_2O \underset{k_\ell}{\overset{k_r}{\rightleftarrows}} H_3O^+ + A^- \qquad K = \frac{k_r}{k_\ell}$$

The equilibrium constant depends on parameters such as pressure and temperature. Assume, for example, that the temperature can be abruptly altered in (say) a few microseconds. At the very moment of this disturbance the system can no longer be considered at equilibrium, and it will begin to approach a new equilibrium situation. This process is called relaxation. Let the concentrations of HA, H_3O^+, and A^- be c_{HA}, $c_{H_3O^+}$, and c_{A^-}, respectively. When the new equilibrium is attained these concentrations are $c_{HA}^{(e)}$, $c_{H_3O^+}^{(e)}$, and $c_{A^-}^{(e)}$, respectively. The rate of reaction to the right is

$$\frac{dc}{dt} = k_r c_{HA} - k_\ell c_{H_3O^+} c_{A^-}$$

$$\frac{d(\Delta c)}{dt} = k_r \left(c_{HA}^{(e)} - \Delta c \right) - k_\ell \left(c_{H_3O^+}^{(e)} + \Delta c \right) \left(c_{A^-}^{(e)} + \Delta c \right)$$

If Δc represents a deviation from equilibrium concentration,

$$c_{HA} = c_{HA}^{(e)} - \Delta c$$

$$c_{H_3O^+} = c_{H_3O^+}^{(e)} + \Delta c$$

$$c_{A^-} = c_{A^-}^{(e)} + \Delta c$$

The product of the latter two terms is

$$c_{H_3O^+}^{(e)} c_{A^-}^{(e)} + \Delta c \left(c_{H_3O^+}^{(e)} + c_{A^-}^{(e)} \right)$$

when $(\Delta c)^2$ is neglected. This is permitted as long as the extent of deviation from the equilibrium position is small. In the equilibrium situation,

$$\frac{dc}{dt} = k_r c_{HA}^{(e)} - k_\ell c_{H_3O^+}^{(e)} c_{A^-}^{(e)} = 0$$

so that

$$\frac{d(\Delta c)}{dt} = \underbrace{\left[-k_r - k_\ell\left(c^{(e)}_{H_3O^+} + c^{(e)}_{A^-}\right)\right]}_{\text{abbreviated as } -1/\tau} \Delta c$$

Just as for a first-order equation, after integration

$$\Delta c_t = \Delta c_{t=0} \exp - \frac{t}{\tau} \qquad \text{or} \qquad \ln \Delta c_t = \ln \Delta c_{t=0} - \frac{t}{\tau}$$

The parameter τ, which can be calculated from the relationship between concentration and time, is called the relaxation time. It is equal to the time in which Δc decreases to $(\Delta c)/e$. Plotting $\ln(\Delta c)$ versus time affords a straight line of slope $-1/\tau$.

By using more than one value for $c^{(e)}_{H_3O^+} + c^{(e)}_{A^-}$, k_r and k_ℓ can be calculated from measurements of the relaxation time.

TEMPERATURE JUMP

One of the methods commonly used is a temperature jump technique. A temperature rise of 2 to 10°C is attained in ca. 1 μs by passing an electric pulse through the solution. This is usually accomplished by discharging a condenser.

Figure 6.3 depicts an instrument where the change of equilibrium position is detected by ultraviolet spectroscopy. In principle, other techniques can also be used, for example, conductivity. The screen of

For more information, see E. Caldin, Chem. Brit. 11, no. 1:4 (1975), and G. G. Hammes in Investigation of Rates and Mechanisms of Reactions, Part II, G. G. Hammes, ed., (Wiley-Interscience, New York, N.Y., 1974), Chap. 4.

the oscilloscope displays a curve (Fig. 6.4) in accordance with the equation

$$\Delta c_t = \Delta c_{t=0} \exp \frac{-t}{\tau}$$

The relaxation time can be calculated directly from the curve. A wide range of rates can be determined corresponding to reaction times from about 10 μs to 1 s. Consecutive reactions can often be separated by simply changing the time base of the oscilloscope. This possibility is extremely useful in, e.g., unraveling the sequence of steps in enzymatic processes.

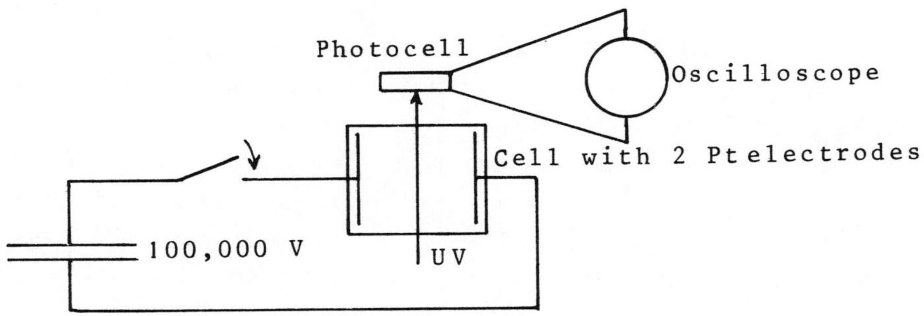

FIG. 6.3

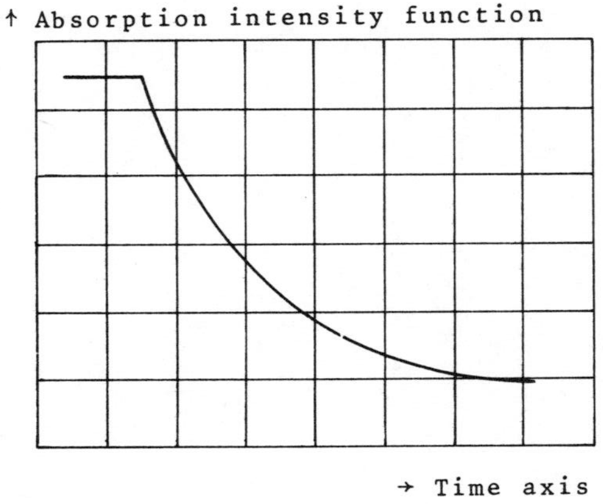

FIG. 6.4

A. Weissberger, Technique of Organic Chemistry, Vol. VIII, Part 2 (Wiley-Interscience, New York, 1963), p. 977.

ELECTRICAL FIELD

In a practical example a condenser of 0.02 μF was charged with 100,000 V. The energy of this pulse is expressed by

$$E = \frac{1}{2} CV^2$$

where $C = 2 \times 10^{-8}$ farad, $V = 10^5$ volt, and $E = 1/2 \times 2 \times 10^{-8} \times 10^{10} = 10^2$ J. The volume of the cell is 1 cm^3. Assuming that the cell absorbs all the energy, the temperature rise would be 25°C. Because of energy losses, the actual rise is only about 6°C. The electrical discharge brings about electrolysis. However, the number of moles converted is small. The amount of electricity is $Q = CV$ or $2 \times 10^{-8} \times 10^5 = 2 \times 10^{-3}$ C. Per equivalent conversion, 1 faraday (= 96,500 C) is necessary. The number of equivalents converted is $(2 \times 10^{-3})/(96 \times 10^3) \approx 2 \times 10^{-8}$. By using a salt concentration of 0.1 mol/liter and 0.1 mmol in the cell, the fraction converted is $(2 \times 10^{-8})/10^{-4} = 2 \times 10^{-4}$, i.e., a small amount. In other words, it is not the amount of charge but the high voltage which is important.

Farad = coulomb/volt, and coulomb = ampere × second. Thus, farad × volt^2 = A × s × V, which equals watt × second = joule.

1 cm^3 at 25° requires 25 cal = 105 J.

The temperature can also be increased by absorption of a short laser or microwave pulse. The latter method has been used for determination of the rate constants in the dissociation equilibria of water and heavy water. At 25°C,

D. H. Turner, G. W. Flynn, N. Sutin, and J. V. Beitz, J. Amer. Chem. Soc. 94:1554 (1972).

G. Ertl and H. Gerischer, Z. Elektrochem. 66:560 (1962).

	For H_2O	For D_2O
k_r (liter mol^{-1} s^{-1})	14.3×10^{10}	8.4×10^{10}
k_1 (s^{-1})	25.6×10^{-6}	2.5×10^{-6}

$$H^+ + OH^- \underset{k_1}{\overset{k_r}{\rightleftarrows}} H_2O$$

$$D^+ + OD^- \underset{k_1}{\overset{k_r}{\rightleftarrows}} D_2O$$

Eigen and Schoen measured the relaxation time for the equilibrium

$$CH_3COOH + H_2O \underset{k_{-1}}{\overset{k_r}{\rightleftarrows}} CH_3COO^- + H_3O^+$$

In this case the disturbance of the equilibrium was obtained by the application of a high-voltage electric field. The equilibrium is displaced to the right when the ions in the ion pairs H^+A^- are dissociated under the influence of the strong external field:

$$HA \rightleftarrows H^+A^- \rightleftarrows H^+ + A^-$$

When the field is applied, the dissociation of the acetic acid is found to increase as measured conductometrically. The increase in conductance, indicative of increased dissociation, is not instantaneous. See Fig. 6.5. The oscilloscope trace of the change in the conductivity permits calculation of τ as well as k_r and k_1 as in the previous approach. For acetic acid at 20°C, $k_r = 8 \times 10^5\ s^{-1}$ and $k_1 = 4.5 \times 10^{10}$ liter mol^{-1} s^{-1}, both approximately ±10%. The applied electric field was of the order of 100 kV/cm.

The rate constants calculated for the reactions $H_3O^+ + A^-$, $H_3O^+ + OH^-$, and $D_3O^+ + OD^-$ are very large. This is often true for ion recombination reactions. The value of the bimolecular rate constant 14.3×10^{10} liter $mol^{-1}\ s^{-1}$ for $H_3O^+ + OH^-$ is the highest rate constant ever measured for a second-order reaction in solution at room temperature. These rate constants are approximately equal to the value calcu-

Manfred Eigen (Nobel Prize, 1967); M. Eigen and J. Schoen, Z. Elektrochem. 59:483 (1955).

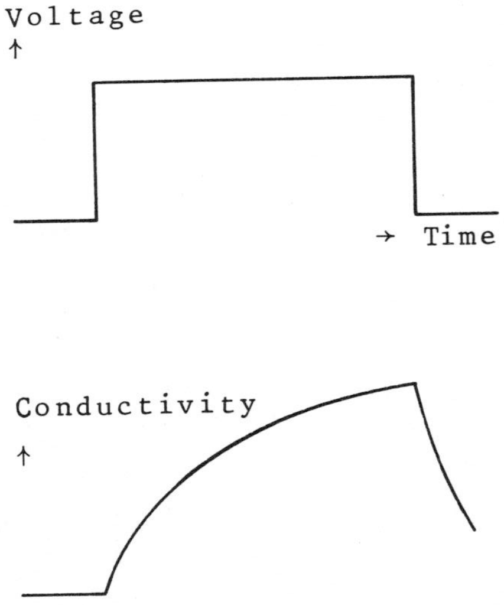

FIG. 6.5

"How fast is fast?": Eyring.

H. Eyring and E. M. Eyring, Modern Chemical Kinetics (Reinhold, New York, 1963), p. 85.

lated on the assumption that each collision between a positive and negative ion is fruitful. In this case where diffusion is the only rate-limiting factor, the expression "diffusion controlled" is used. For the diffusion-controlled reaction

$$H_3O^+ + OH^- \longrightarrow 2H_2O$$

the calculated rate constant, $k = 13 \times 10^{10}$ liter mol^{-1} s^{-1}, is approximately equal to the experimental value.

ULTRASONICS

In all these instances the change of environment occurred discontinuously. However, it is also possible to arrange that such changes take place continuously. For example, the pressure and the temperature of a system change simultaneously when a sound wave of frequency ν is beamed through it. See Fig. 6.6. A sound wave is longitudinal. When this wave runs through the medium, a periodic disturbance takes place whose frequency is also equal to ν. Assume a liquid system in which equilibrium exists between compounds A and B:

$$A \underset{k_{-1}}{\overset{k_1}{\rightleftarrows}} B$$

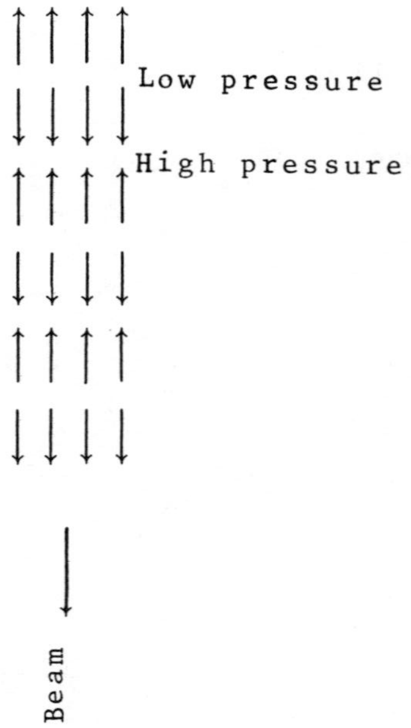

FIG. 6.6 Sound wave.

J. Lamb, Z. Elektrochem. 64:135 (1960).

Generally, pressure and temperature fluctuations disturb the position of equilibrium. If the frequency of the sound wave is very high, the pressure and temperature fluctuations are so fast that equilibrium lags behind, and little shift in equilibrium occurs. At lower frequency, the equilibrium will coordinate well with the temperature and pressure changes.

There are now two sets of periodic changes:

1. The pressure and temperature changes induced by the sound waves
2. The fluctuations of the equilibrium position

These alterations have the same frequency ν but differ in phase. Depending on this difference in phase, absorption of sound energy takes place. The maximum is reached for the frequency ν' at which point $1/\tau = 2\pi\nu'$. See Fig. 6.7.

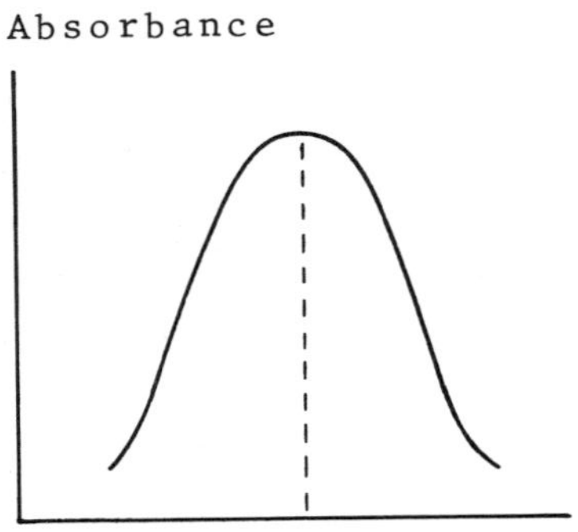

FIG. 6.7 ν' in the region of ultrasonic waves: $1/\tau = 2\pi\nu'$.

Let the concentration of A and B in the above-mentioned equilibrium be a - x and x, respectively:

$$\frac{dx}{dt} = k_1(a - x) - k_{-1}x$$

Let $\Delta x = x - x^{(e)}$, where $x^{(e)}$ is the equilibrium value of x:

$$\frac{d(\Delta x)}{dt} = k_1(a - x^{(e)}) - k_1\,\Delta x - k_{-1}x^{(e)} - k_{-1}\,\Delta x$$

Since at equilibrium

$$k_1(a - x^{(e)}) = k_{-1}x^{(e)}$$

then

$$\frac{d(\Delta x)}{dt} = -(k_1 + k_{-1})\ \Delta x = -\frac{1}{\tau}\ \Delta x$$

If, moreover, $K = k_1/k_{-1}$ is known, then k_1 and k_{-1} can be calculated separately. Sometimes $K << 1$, in which case $k_1 + k_{-1} \simeq k_{-1}$.

Examples: In ultrasonic sound experiments with methylcyclohexane at 16°C, $1/\tau = 8.8 \times 10^5\ s^{-1}$. No sound absorption takes place when this experiment is done with cyclohexane. Its two conformations are equally probable, and thus the equilibrium constant is always unity; the equilibrium position does not shift when the temperature changes. The same is true for 1,1-, cis-1,2-, and cis-1,4-dimethylcyclohexane. The latter compounds, however, can be studied with nuclear magnetic resonance.

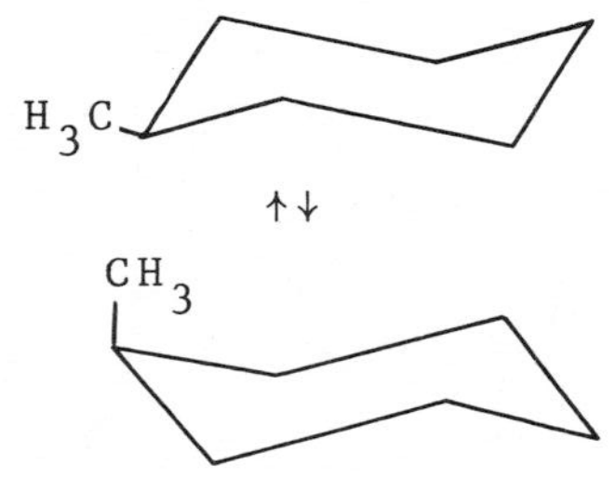

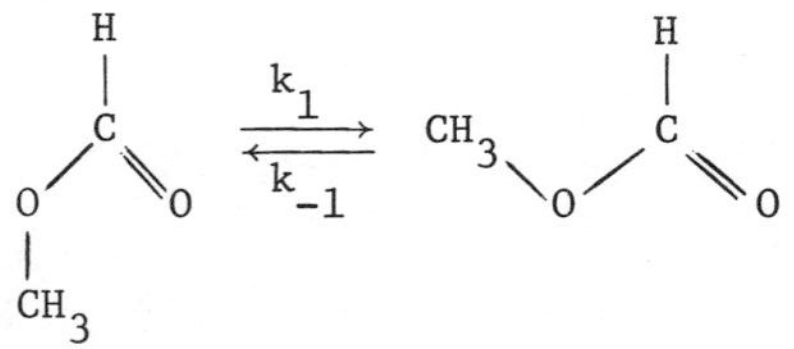

Methyl formate:	$k_1 + k_{-1} = 2.2 \times 10^6\ s^{-1}$	at 25°C
Acrolein:	$k_1 + k_{-1} = 1.1 \times 10^9\ s^{-1}$	at 25°C
Methyl vinyl ether:	$k_1 + k_{-1} = 1.4 \times 10^9\ s^{-1}$	at -25°C

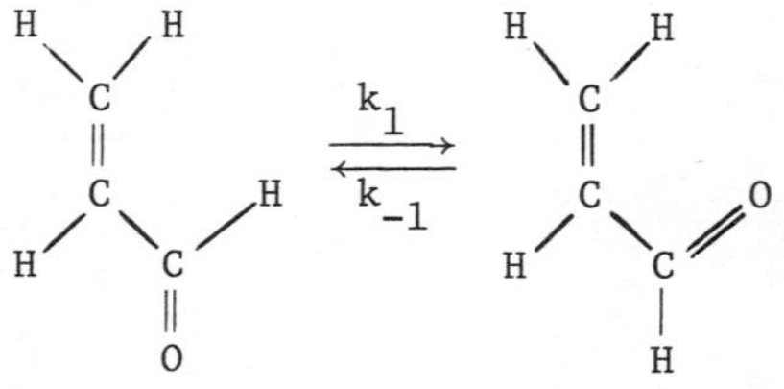

S-trans S-cis

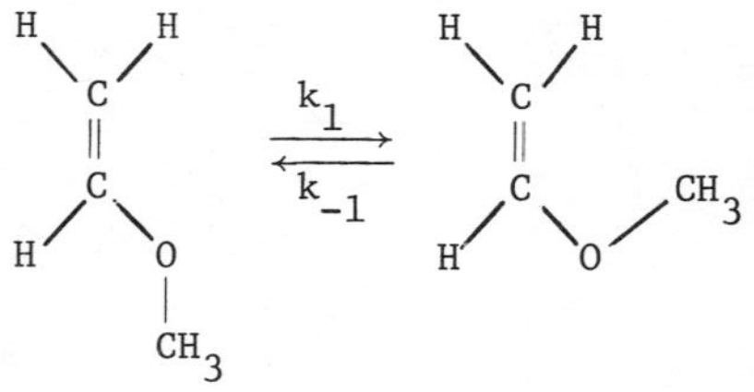

DYNAMIC NUCLEAR MAGNETIC RESONANCE (NMR) SPECTROSCOPY

This technique will be illustrated by considering acetylenic hydrogen exchange. The determination of the rate constant for acetylenic hydrogen exchange in tert-butylethyne, $(CH_3)_3CC{\equiv}CH$, has been mentioned on page 61.

Two peaks appear, one for each kind of hydrogen. See Fig. 6.7. The compound was dissolved in acetone-D_2O, which also contains a small amount of the base KOD. The ≡CH peak gradually became smaller by substitution of deuterium for hydrogen. The disappearance rate of the ≡CH is equal to the ionization rate:

$$RC{\equiv}CH + OD^- \rightleftharpoons RC{\equiv}C^- + HOD$$

The reaction can be measured by this method only when the rate is not too rapid. When the reaction has a half-life shorter than approximately 10 s, the NMR technique can still be applied, but in a different way. For example, for $(CH_3)_3CSC{\equiv}CH$ deuterium exchange is much more rapid. A solution of this compound in a neutral mixture of acetone and 10 vol % water shows sharp peaks for tert-butyl and ≡C—H, apart from the acetone and water signals.

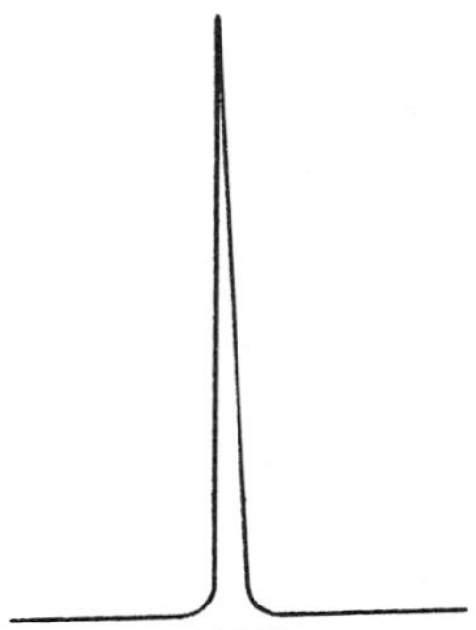

Addition of a small amount of KOH to the solution produces broadening of the peak corresponding to the acetylenic hydrogen. The broadening is related to the equilibrium

$$RS{-}C{\equiv}CH + OH^- \rightleftharpoons RS{-}C{\equiv}C^- + H_2O$$

involving rapid hydrogen exchange between acetylene and water catalyzed by base. What is the relation between exchange rate and peak width? In this connection it is important to know how the spectrum arises. The behavior of the hydrogen nucleus can be related to that of a magnet. This magnetic behavior results from the nucleus having a spin angular moment. The magnetic moment μ is proportional to the spin quantum number I and can be represented by a vector:

Magnetic moment $\mu = \gamma I$ γ = gyromagnetic ratio

If such a magnet, the proton, is placed in a strong, homogeneous magnetic field, it can orient itself in two ways, aligned or opposed. The projection of μ in the direction of the field strength is either $+\gamma(1/2)(h/2\pi)$ or $-\gamma(1/2)(h/2\pi)$. The energy of (1) is slightly lower than the energy of (2). There are slightly more protons in the lower than in the higher level. The ratio of the populations is given by the Boltzmann equation $n_2/n_1 = \exp(-\Delta E/RT)$, where ΔE, the energy difference per mole, is extremely small. For the usual magnetic strength of 23,500 G (gauss), $\Delta E \simeq 0.042$ J/mol. Then, at room temperature, $n_2/n_1 = \exp(-\Delta E/RT) \simeq 0.99998$. The equilibrium ratio can be established, because transitions take place between both levels under the influence of the fluctuating magnetic fields due to the fluctuating lattice motions of the nuclei which comprise the neighboring molecules. The average lifetime of a particle in one of the levels is given by the symbol T. In other words, the rate constant of the particles from one to the other level is 1/T.

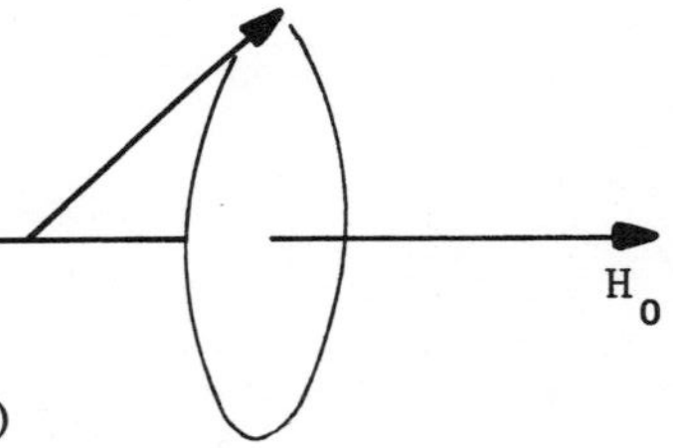

(1)

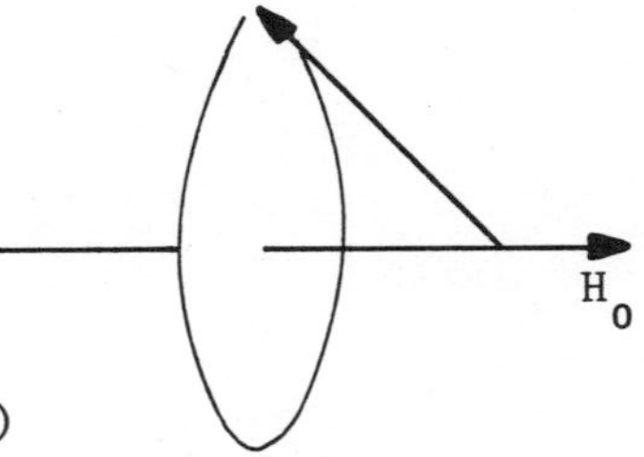

(2)

Just as in other spectroscopic experiments, the transition can also be brought about through absorption of electromagnetic radiation with a frequency ν, so that $\Delta E = h\nu$, where h is Planck's constant. What frequency is required? When $\Delta E = 0.042$ J/mol, the energy difference per molecule is

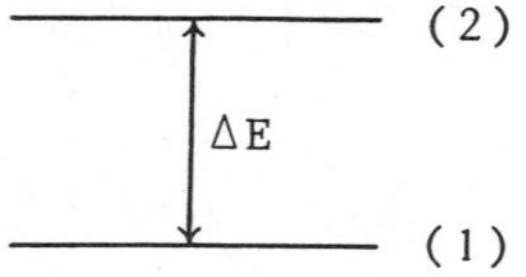

$$\Delta\varepsilon = \frac{0.042}{6 \times 10^{23}} \text{ J} \quad \text{and} \quad \nu = \frac{\Delta\varepsilon}{h} \simeq 10^8 \text{ s}^{-1} = 100 \text{ MHz}$$

$h = 6.6256 \times 10^{-34}$ Js

The corresponding wavelength is

$$\lambda = \frac{3 \times 10^{10}}{10^8} = 300 \text{ cm}$$

This wavelength is in the range of radio waves.

The radiation of 100 MHz causes transition from the lower to the higher as well as from the higher to the lower level. Since the population is a little bit greater in the lower level than in the higher one, more nuclei transfer from the lower to the higher level than vice versa. The net result is a transition from lower to higher level. This transition requires energy, and therefore radiation energy will be absorbed. A recorder or an oscilloscope visualizes this absorption of energy. Owing to the absorption process, the ratio n_2/n_1 comes closer to 1 than the original value 0.99998. Under these circumstances the equilibrium is perturbed, and the system will try to restore the equilibrium. Just as on page 146 this behavior is called relaxation. The relaxation curve is given by the expression

$$\frac{n_2}{n_1} = \left(\frac{n_2}{n_1}\right)_0 \exp \frac{-t}{T}$$

The relaxation time is now called T. It is related to the breadth of the absorption line. This agrees with the Heisenberg Uncertainty Principle expressed in the form $\Delta E \Delta t = h/2\pi$. In this case $\Delta t = T$, the lifetime of a proton in a magnetic state. The uncertainty in the energy transfer, $h\,\Delta\nu$, is about twice that in each level.

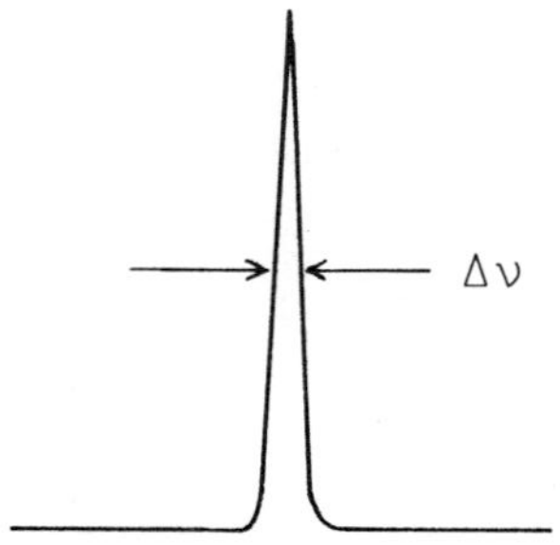

Thus, $h\ \Delta\nu = (h/\pi)(1/T)$ or $\pi\ \Delta\nu = 1/T$, where $\Delta\nu$ is the width of the peak at half-height and T is the lifetime of a particle in a magnetic state. In this case the particle is a proton, for example, the acetylenic proton of $(CH_3)_3CSC{\equiv}CH$. After a lifetime T the spin reverses, or, in other words, the rate constant of spin reversal is 1/T:

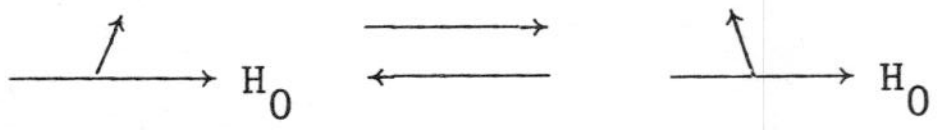

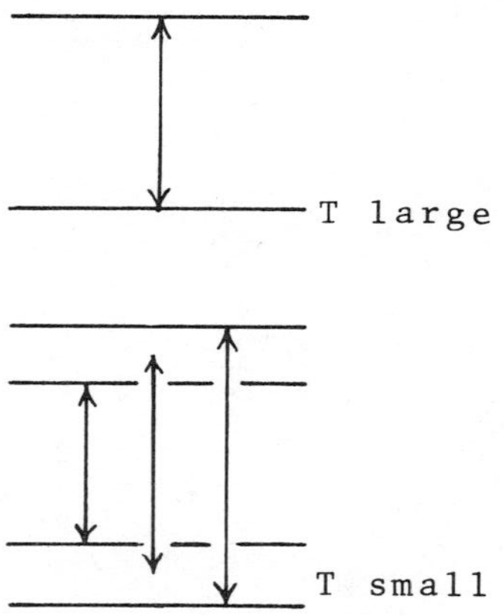

Now, assume this compound takes part in the chemical reaction

$$(CH_3)_3CSC{\equiv}CH + OH^- \rightleftharpoons (CH_3)_3CSC{\equiv}C^- + H_2O$$

The spin state of the acetylenic hydrogen is altered in two distinct processes:

1. The spin reversal, rate constant 1/T
2. The chemical exchange, first-order rate constant k

The total rate constant is k + 1/T, which gives rise to a new relaxation time T' such that 1/T' = k + 1/T (Figs. 6.8 and 6.9):

Without exchange: $\frac{1}{T} = \pi\ \Delta\nu$

With exchange: $\frac{1}{T'} = k + \frac{1}{T} = \pi\ \Delta\nu'$

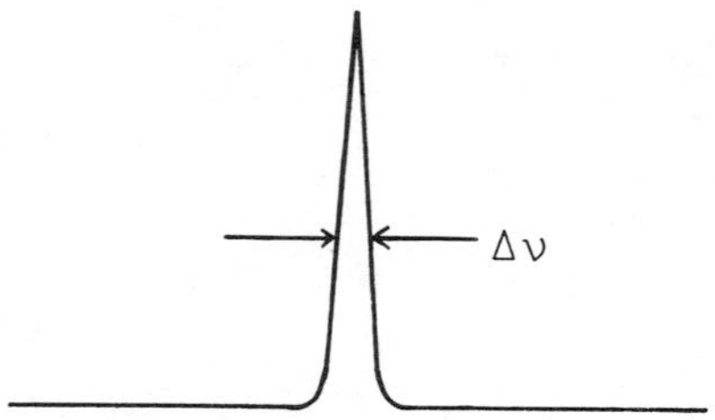

FIG. 6.8

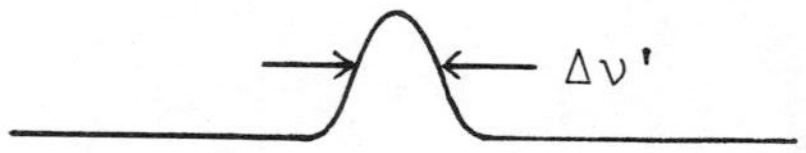

FIG. 6.9

The rate constant can be calculated from the line broadening. In acetone +10 vol % H_2O at 25°C, the first-order rate constant in the equation $v = k_1 c_{\equiv CH}$ amounts to $0.91\ s^{-1}$ for $c_{OH^-} = 0.001$ mol/liter. The half-life is $t_{1/2} = (\ln 2)/0.91 = 0.8$ s.

In the NMR spectrum of a mixture of extremely pure alcohol and water, different peaks appear for the hydroxyl hydrogen of alcohol and of water. See Fig. 6.10. Addition of a trace of base or acid to the mixture causes the exchange rate of the hydroxyl protons to increase due to reactions such as

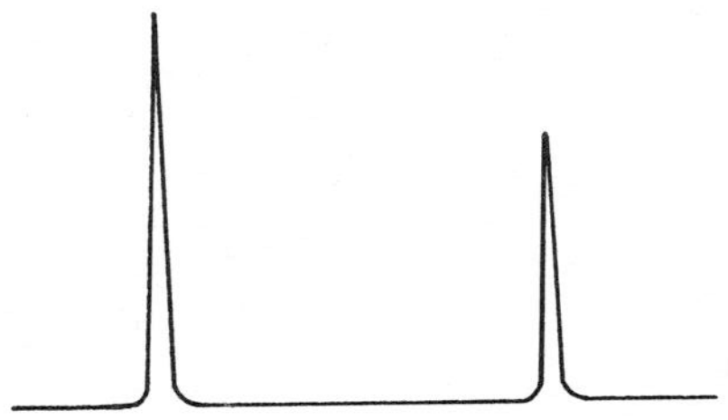

FIG. 6.10

$$C_2H_5OH + OH^- \underset{k_{-2}}{\overset{k_2}{\rightleftharpoons}} C_2H_5O^- + H_2O$$

$$C_2H_5OH + H_3O^+ \underset{k'_{-2}}{\overset{k'_2}{\rightleftharpoons}} C_2H_5\overset{+}{O}H_2 + H_2O$$

The bands are broadened (Fig. 6.11), and it can be calculated from the extent of broadening that in alcohol with 11 weight % water at 22°C the rate constants are

FIG. 6.11

$$k_2 = 2.6 \times 10^6 \text{ liter mol}^{-1}\ s^{-1}$$

and

$$k'_2 = 2.8 \times 10^6 \text{ liter mol}^{-1}\ s^{-1}$$

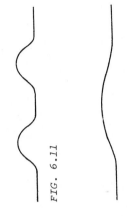

FIG. 6.12

With more acid or base the bands are still broader because the exchange rate is faster (see Fig. 6.12). Moreover, the band positions move in each other's direction. With increasing exchange rates the bands coincide. In popular terms, the exchange of protons between alcohol and water has become so rapid that the NMR instrument can no longer follow the reactions. The NMR spectrum only shows the average position of the hydroxyl hydrogen. When very rapid exchange occurs, there is one sharp peak; see Fig. 6.13.

The shape of an absorption band can be calculated when the chemical shifts, coupling constants, and relaxation times are known. By comparing the shape of the bands simulated by computer for different relaxation times with the experimentally observed line shape, the actual relaxation times and, consequently, the rate constants can be obtained.

The central N—C bond of N,N-dimethyl formamide has some double-bond character. As a consequence, the rotation around this bond is relatively slow. The methyl groups are in different surroundings and therefore display different chemical shifts in the spectrum. With rising temperature the peaks broaden because the rotation about C—N and therefore the exchange of environments, is accelerated. At 123°C the peaks coincide. When the temperature is increased still further, this single methyl peak sharpens. At 170°C there is one sharp peak at a position corresponding to the exact center between the original peaks. At this temperature and according to the time scale of ^{1}H NMR, the methyls have become equivalent, and we may speak of free rotation about the C—N bond.

The relaxation times of this system were determined by comparing experimental and calculated line shapes (see above). The increase in relaxation time is related to the rate constant in the equation $k = 1/T - 1/T'$ (see p. 158). For example, at 100°C the rate constant

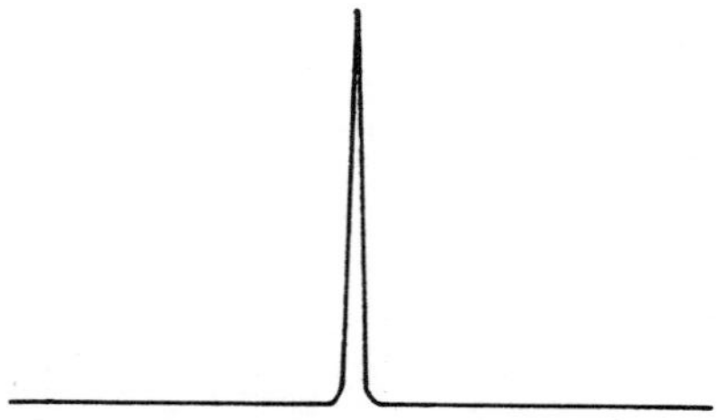

FIG. 6.13

"Total line shape analysis"

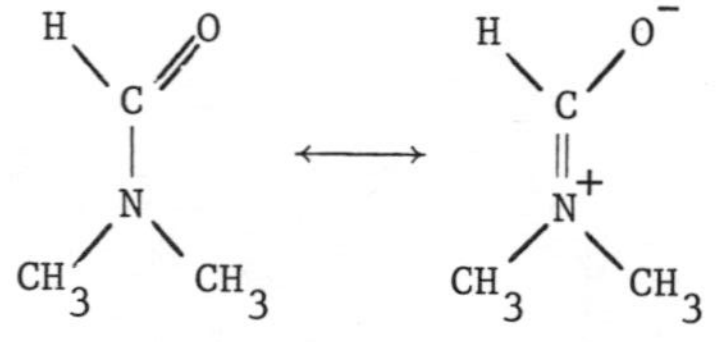

F. Conti and W. von Philipsborn, Helv. Chim. Acta 50:603 (1967).

is 10 s^{-1}. The log of the rate constant at temperatures varying from 100 to 130° was plotted in Arrhenius fashion. The slope of this linear plot corresponded to an activation energy of 92 kJ/mol.

In cyclohexane the axial and equatorial hydrogen atoms interconvert. Axial and equatorial hydrogen atoms have different environments in the molecule and therefore also different peaks in the NMR spectrum, at least when the temperature is sufficiently low and the rate of interconversion decreases to a value of the order of 100 interconversions per second. The rate constant of the interconversion can be calculated from the shape of the peaks. The activation energy follows from the temperature dependency of the rate constant. The activation energy is some 40 kJ/mol.

R. K. Harris and N. Sheppard, J. Mol. Spectrosc. 23:231 (1967).

Range of different methods:

Log half-lifetime in s: 1 0 -1 -2 -3 -4 -5 -6 -7 -8 -9

Classical ⟶

Stopped flow ⟶

Temperature jump ⟶

NMR ⟶

⟵ Ultrasound ⟶

Flash photolysis ⟶

D. N. Hague, Fast Reactions (Wiley-Interscience, London, 1971), p. 143.

7 Catalysis

Emphasis on acid-base catalysis

DEFINITION

Catalysis has been defined in several ways. A kinetically interesting definition has been given by R. P. Bell: "A substance is said to be a catalyst for a reaction in a homogeneous system when its concentration occurs in the velocity expression to a higher power than it does in the stoichiometric equation."

R. P. Bell, Acid-Base Catalysis (Cornell Univ. Press, Ithaca, N. Y., 1941), p. 3.

$$RCOOR' + H_2O \longrightarrow \cdots$$

$$v = k[RCOOR'][H_3O^+]$$

H_3O^+ is a catalyst that does not appear in the stoichiometric equation.

Acceleration of the reaction is equivalent to a lowering of the energy barrier between the initial state and product. The catalyst does not change the energy levels of the initial state and product. The equilibrium position does not change. In this connection it must be emphasized that each system progresses toward its equilibrium

$$A + B \rightleftarrows C + D$$

position. This position is determined by the difference in Gibbs function between A + B and C + D. When the C + D level is much below the A + B level, the reaction to the right is driven nearly to completion. The arrow to the left is generally omitted. In the reverse case, the C + D level being appreciably higher, the reaction to the right is almost negligible except when C + D takes part in a subsequent reaction which removes these components of the equilibrium. There is no catalyst to change this behavior. An example is the reaction

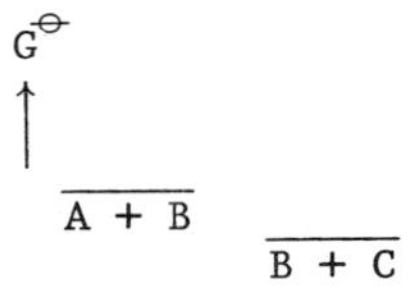

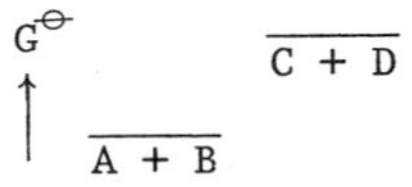

$$CH_4 + CO_2 \longrightarrow CH_3OH + CO$$

M. Prettre, Catalysis and Catalysts (Dover, New York, 1963), p. 10ff.

The answer to the question as to whether this reaction could occur can be obtained from the Gibbs function. At a temperature of T (in K),

K = ° Kelvin

$$\Delta G^{\ominus} = 157{,}640 - 40.6 \times T \text{ J/mol}$$

Thus, at ambient temperatures (T = 300 K), $\Delta G^{\ominus}$ = 145,460 J/mol, and at the industrially very high temperature of 1500 K, $\Delta G^{\ominus}$ = 96,740 J/mol. The equilibrium constant is equal to $\exp[-\Delta G^{\ominus}/RT]$ and at T = 300 K is $\exp(-58) = 10^{-25}$ and at T = 1500 K is $\exp(-7.7) = 10^{-3.3}$. At each temperature the equilibria are displaced almost completely to the left. Normally, searching for a catalyst which would promote the reaction to the right would seem to be unpromising. From a thermodynamic viewpoint, the reaction to the left might occur most readily. However, when it is slow it does make good sense to search for a catalyst.

$R = 8.3143 \text{ J mol}^{-1} \text{ K}^{-1}$

HOMOGENEOUS CATALYSIS

In industry the great majority of chemical processes are performed with one catalyst or another. Most of these catalysts are heterogeneous. Heterogeneous catalysis involves adsorption processes and crystal structures of catalysts and does not fall within the scope of this discussion, which is restricted to homogeneous catalysis. Such reactions take place in one phase, which, in our examples, will always be the liquid phase. Often, the catalysis is due to one of the following effects:

1. Activation of one of the reagents
2. Combination of two reagents in one ligand sphere
3. Elimination of symmetry restrictions

An example of the first effect is the case of H_3O^+ catalysis where proton addition increases the positive charge of a molecule and thereby facilitates attack by a nucleophile.

An example of the second effect often occurs with the use of transition metals as catalysts, where, besides activation, the adjacency of the reagents in the coordination sphere of the metal often plays an important role in determining the extent of catalysis. By way of illustration, the industrially important conversion of olefins into aldehydes, called hydroformylation, is catalyzed by cobalt compounds:

See, e.g., R. F. Heck and D. S. Breslow, J. Amer. Chem. Soc. 83:4023 (1961).

$$RCH{=}CH_2 + CO + H_2 \longrightarrow RCH_2CH_2CHO$$

Pressures of 100 atm and temperatures of approximately 100°C are often required. Under the reaction conditions the cobalt is converted into its dissociated carbonyl complex:

$$HCo(CO_4) \rightleftharpoons HCo(CO)_3 + CO$$

The tricarbonyl has an open ligand position and is able to bind an olefin molecule:

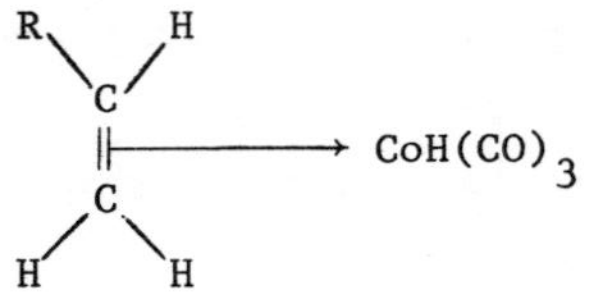

Subsequently, a sequence of reactions occurring in the coordination sphere of the cobalt leads to product and reconstitution of the effective catalyst complex $HCo(CO)_3$:

$$\text{Complex} \longrightarrow RCH_2CH_2Co(CO)_3 \xrightarrow{CO} RCH_2CH_2Co(CO)_4 \longrightarrow$$

$$RCH_2CH_2COCo(CO)_3 \xrightarrow{H_2} RCH_2CH_2CHO + HCo(CO)_3$$

The rate of the reaction approximately fits the equation

$$\frac{d[\text{aldehyde}]}{dt} = k[\text{olefin}][\text{Co}]p_{H_2}p_{CO}^{-1}$$

Nickel acetylacetonate, $Ni(Acac)_2$, is catalytically active in the polymerization of isocyanides:

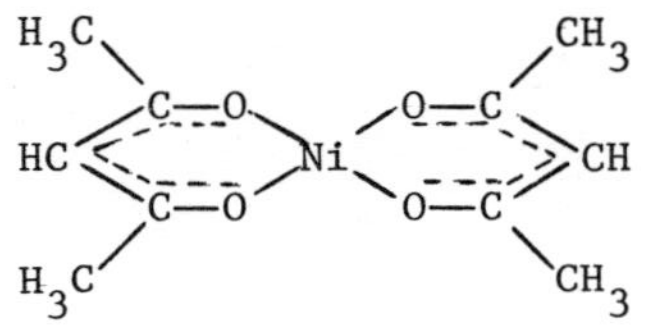

$$n\ R{-}N{=}C \xrightarrow{Ni(Acac)_2} (R{-}N{=}C\lt)_n$$

The rate of polymerization is proportional to the square root of the catalyst concentration, $[Ni(Acac)_2]^{1/2}$. The half power indicates that the catalyst dissociates. The polymerization probably takes place through catalysis by the dissociated species, $Ni(Acac)^+$.

An example of the third catalytic effect is the elimination of symmetry restrictions in the isomerization of quadricyclane to norbornadiene. See Fig. 7.1. Without a catalyst, the half-life of this isomerization is larger than 14 h at 140°C. The reaction is to be viewed as the reverse of a typical thermal [2 + 2] cycloaddition. According to the Woodward-Hoffmann rules of orbital symmetry conservation, a synchronous reaction mechanism of this type is symmetry forbidden. The "forbidden character" is eliminated by addition of a rhodium, palladium, or platinum complex. For instance, with a certain dirhodium complex the half-life is reduced to 45 min at -26°C.

Different reasons have been given for this accelerating effect. One viewpoint favors the much greater choice of filled and empty orbitals of different symmetries provided by the transition metal. Another possibility is elimination of the requirement for concertedness of the ring opening induced by the presence of an organometallic intermediate. Some theoreticians believe that the acceleration is not the "removal of forbiddenness" but the lowering of high-energy orbitals of the organic compound by configuration interaction with orbitals of the metal.

$$Ni(Acac)_2 \rightleftharpoons {}^{+}NiAcac + Acac^{-}$$

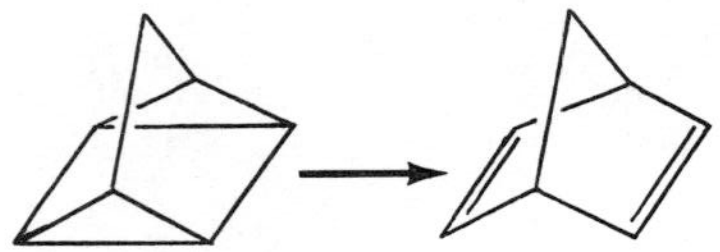

FIG. 7.1

H. Hogeveen and H. C. Volger, J. Amer. Chem. Soc. 89:2486 (1967). See also Tetrahedron Lett. 1973:3667.

Acids and bases are defined in different ways. We shall limit the discussion to the properties defined by Brønsted: An acid is a compound capable of donating a proton; a base is a compound capable of binding a proton. There is no restriction on the charge types of acids and bases.

In organic chemistry a large number of reactions occur only in the presence of an acid, e.g., the formation and cleavage of acetals, the pinacol, and the benzidine rearrangements. Other reactions are catalyzed by bases, e.g., the aldol condensation and the benzilic acid rearrangement. A number of reactions are catalyzed by acids as well as bases, e.g., the hydrolysis of esters and the aldol condensation.

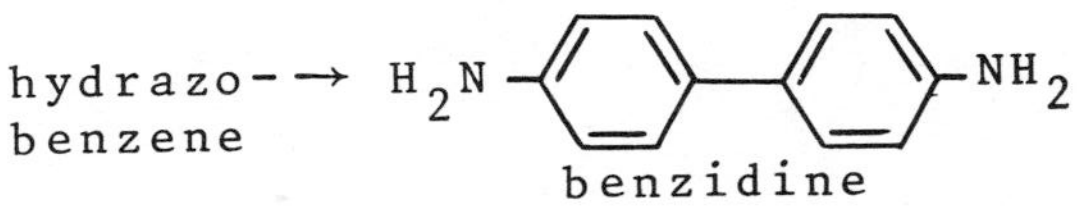

An example of a reaction catalyzed by hydroxyl ions is the decomposition of diacetone alcohol. The rate can be followed dilatometrically in water as solvent. The reaction appears to be first order in the concentration of diacetone alcohol (see Fig. 7.3):

C. C. French, J. Amer. Chem. Soc. 51:3215 (1929).

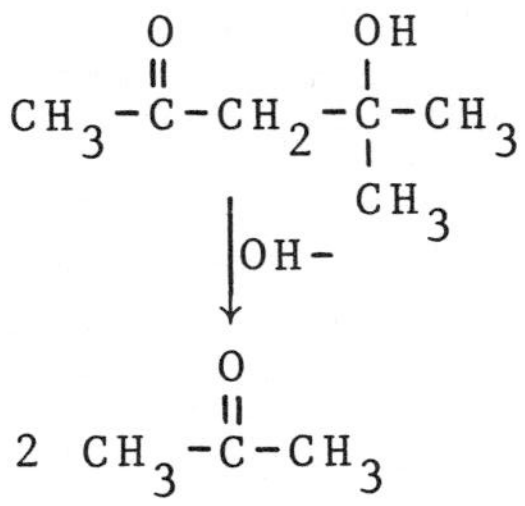

$$v = k_1 c_{\text{diac. alc.}}$$

The first-order rate constant k_1 was determined at various concentrations of base; see Table 7.1. These data fit the equation $v = k_2 c_{\text{diac. alc.}} c_{OH^-}$, in agreement with the mechanism proposed for the reaction. Its first step is

(1) 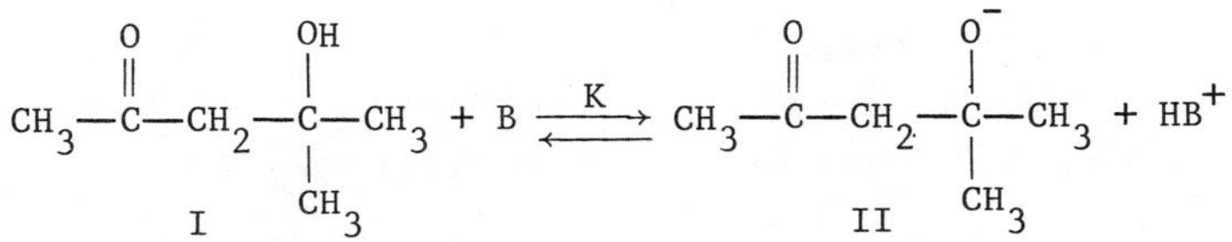

TABLE 7.1

[NaOH] (mol/liter)	$k_1 \times 10^5$ (s^{-1})	$k_2 \times 10^4$ (liters mol^{-1} s^{-1})
0.005	3.87	77
0.010	7.78	78
0.020	15.7	79
0.040	32.0	80
0.100	79.8	80

$$k_2 = \frac{k_1}{c_{OH^-}}$$

This step is a rapid proton transfer from diacetone alcohol to base. The base does not have to be hydroxyl ion. The proton can be accepted by each base of appropriate strength present in the solution, e.g., acetate ion, amine, water, etc.

(2) 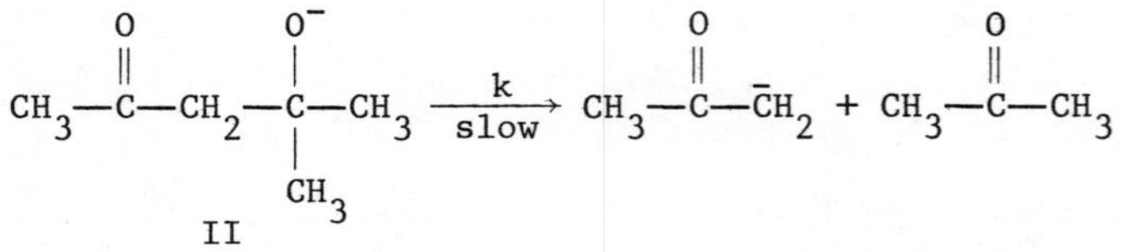

$$CH_3-\overset{O}{\overset{\|}{C}}-CH_2-\overset{\overset{O^-}{|}}{\underset{\underset{CH_3}{|}}{C}}-CH_3 \xrightarrow[\text{slow}]{k} CH_3-\overset{O}{\overset{\|}{C}}-\bar{C}H_2 + CH_3-\overset{O}{\overset{\|}{C}}-CH_3$$

II

(3) 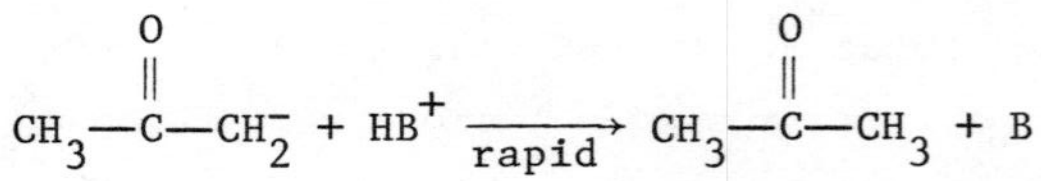

$$CH_3-\overset{O}{\overset{\|}{C}}-CH_2^- + HB^+ \xrightarrow[\text{rapid}]{} CH_3-\overset{O}{\overset{\|}{C}}-CH_3 + B$$

Ultimately, the base is released for further catalytic activities. The rate is determined by the slow step:

$$v = kc_{II}$$

The constant of the equilibrium (1) obeys the equation

$$K = \frac{c_{II}c_{HB^+}}{c_I c_B}$$

Moreover, we must take into account the dissociation equilibrium of the base:

$$B + H_2O \rightleftharpoons HB^+ + OH^-$$

$$K_{base} = \frac{c_{HB^+}c_{OH^-}}{c_B}$$

Combination of both equilibria gives

$$\frac{c_{II}}{c_I c_{OH^-}} = K'$$

where K' is the constant of the equilibrium:

$$I + OH^- \overset{K'}{\rightleftharpoons} II + H_2O$$

Substitution of c_{II} in the rate equation leads to

$$v = k'c_I c_{OH^-}$$

This equation is in agreement with the experimental data.

SPECIFIC CATALYSIS

Remarkably, the hydroxyl ion is the only base whose concentration appears in the rate equation. When bases such as acetate of phenolate ions are present in the solution, their concentrations exert little or no influence on the rate. This solution is called specific catalysis by hydroxyl ions.

To prove that the concentrations of bases other than hydroxyl ions indeed do not appear in the rate equation, the rate was measured in a series of phenolate buffers:

$k_1 \times 10^5$ (s^{-1})

0.079	0.078	0.077
0.2 mol/liter of NaOPh	0.15 mol/liter of NaOPh	0.1 mol/liter of NaOPh
0.2 mol/liter of HOPh	0.15 mol/liter of HOPh	0.1 mol/liter of HOPh
	0.05 mol/liter of NaCl	0.1 mol/liter of NaCl

In each solution the concentrations of sodium phenolate and phenol are equal. Sodium chloride was added in order to keep the ionic strength at a constant value. Under these conditions the pH's of the solutions have the same value. Consequently, there is a small but constant hydroxyl concentration. In the three solutions the rate does not deviate from constancy by more than the experimental error. According to this experiment, therefore, the rate does not depend on the total concentration of phenolate ion.

GENERAL CATALYSIS

The counterpart of specific catalysis is *general catalysis*. A reaction is subject to general base catalysis when the concentrations of bases other than hydroxyl ions contribute to the rate. General base catalysis was first observed by Brønsted in the decomposition of nitramide. The reaction, followed by measuring the amount of N_2O evolved, was performed in aqueous solution. When the solution is basic the reaction is almost instantaneous, reflecting the large catalytic coefficient of hydroxyl ions. To keep the hydroxyl ion concentration low, the reaction was run in buffered solutions, usually below pH 7, e.g., in an acetate buffer. Under these conditions it was clear that the rate depends on the acetate ion concentration:

$$H_2N—NO_2 \longrightarrow H_2O + N_2O$$

J. N. Brønsted, Z. Phys. Chem. <u>108</u>:185 (1924).

$$v = k_1 c_{NH_2NO_2}$$

with

$$\frac{k_1}{s^{-1}} = 1.45 \times 10^{-5} + 19.3 \times 10^{-3}[c_{OAc^-}/(mol/liter)]$$

Similarly, in benzoate buffers

$$\frac{k_1}{s^{-1}} = 1.45 \times 10^{-5} + 7.4 \times 10^{-3}[c_{C_6H_5CO_2^-}/(mol/liter)]$$

The first term corresponds to catalysis by water acting as a base at constant concentration. The general expression for the rate equation is

$$k_1 = k_{H_2O}c_{H_2O} + k_{OH^-}c_{OH^-} + k_{OAc^-}c_{OAc^-} + \cdots$$

or

$$k_1 = \sum_i k_i c_{B_i}$$

the sum of the contributions of all bases, i.e., the sum of the products of the concentration of each catalytic base and its specific catalytic coefficient k_i. In this example the term $k_{OH^-}c_{OH^-}$ is very large in basic solutions; however, it is undetected at the pH of an acetate or benzoate buffer where c_{OH^-} is too small (despite a large k_{OH^-}).

BRØNSTED CATALYSIS LAW

Brønsted observed that the equation

$$k_i = GK_{B_i}^{\beta} \tag{7.1}$$

holds in the first approximation. This equation is called the Brønsted catalysis law; k_i is the catalytic constant of the base B_i, K_{B_i} is its base strength, G and β are constants, and β has a value between 0 and 1. The stronger the base, the larger its catalytic action. Acetic acid is a slightly weaker acid than benzoic acid. Correspondingly, acetate ions are slightly more basic than benzoate ions. The catalytic constant of water amounts to $(1.45 \times 10^{-5}\ s^{-1})/(56\ mol/liter) = 2.6 \times 10^{-7}$ liter $mol^{-1}\ s^{-1}$. In agreement with the much lower basicity of water this

Concentration of water = 1000/18 ≃ 56 mol/liter

value is several orders of magnitude smaller than the catalytic constants of acetate and benzoate ions.

The first step, which is rate determining, consists of the abstraction of a proton by a molecule of base:

$$NH_2{-}NO_2 + B \longrightarrow NHNO_2^- + HB^+$$

This step is easier the stronger the base, as expressed by Brønsted's catalysis law. In its logarithmic form,

$$\log k_i = \text{constant} + \beta \log K_{B_i} \qquad (7.2)$$

This law is an example, and actually the first one, of the many so-called linear free-energy relations (LFERs).

Other LFERs include Hammett, Taft, Yukawa-Tsuno, Grunwald-Winstein, Swain-Scott, and Edwards.

Usually, Brønsted's law fits the data reasonably well only when bases of the same type are correlated. For instance, in the nitramide reaction Brønsted observed that the catalytic rate constants of carboxylic ions, $RCOO^-$, obeyed the equation

$$k_i = 2.38 \times 10^{-6} K_{B_i}^{0.83}$$

This $K_B = 1/K_{acid}$.

while the data for aniline and ring-substituted anilines of varying pK_B's fit the equation

$$k_i = 6.5 \times 10^{-6} K_{B_i}^{0.75}$$

$$K_B = \frac{c_{aniline.H^+}}{c_{aniline} c_{H^+}}$$

Summarizing, base catalysis is either specific or general. When it is specific, hydroxyl ion is the only base whose concentration occurs in the rate equation, at least in aqueous solution. When pure ethyl alcohol is the solvent, $C_2H_5O^-$ is substituted for HO^-. When the catalysis is general, the rate equation involves the concentration of each base, at least in principle. Moreover, the catalytic effect of a base increases with increasing base strength.

The mechanism of the reaction determines whether we have specific or general catalysis. In our first example, the decomposition of diacetone alcohol, the first step of the reaction is a rapid proton transfer. In such cases the reaction is subject to specific base catalysis. In the second example, the decomposition of nitramide, the first step is a rate-determining proton transfer exhibiting general base catalysis. Analogously, in acid catalysis we can distinguish specific catalysis by H_3O^+ and general catalysis by acids. In the former case, hydronium ions are the only acid species in aqueous solution to appear in the rate equations--more precisely stated, H_3O^+ in aqueous solution and H_2OS^+ in a solvent HOS. When a reaction is general acid catalyzed, its rate equation involves, at least in principle, each acid present in the solution independently of its charge type. The catalytic action of an acid is stronger the higher its acidity. Here also a Brønsted catalysis law applies:

H_3O^+, CH_3COOH, C_6H_5OH, H_2O, $H_2PO_4^-$

$$k_i = GK_{a_i}^{\alpha} \quad \text{or} \quad \log k_i = \text{constant} + \alpha \log K_{a_i} \tag{7.3}$$

See the review by A. J. Kresge, Chem. Soc. Rev. 2:475 (1973).

G and α are constants; α has a value between 0 and 1.

An example is the acid catalyzed addition of water to ethoxypropadiene:

$$H_2C{=}C{=}CHOC_2H_5 + H_2O \longrightarrow H_2C{=}CHCHO + C_2H_5OH$$

Acid:	H_3O^+	$ClCH_2COOH$	CH_3OCH_2COOH	$HCOOH$	$HOCH_2COOH$	CH_3COOH
pK_a:	-1.74	2.87	3.57	3.75	3.83	4.76
k_{HA} (liters mol^{-1} s^{-1})	1.10	0.0783	0.0291	0.0193	0.0190	0.0054

In this case $\alpha = 0.62$.

What is the meaning of this constant α? In the acid dissociation in aqueous solution,

$$HA + H_2O \rightleftharpoons H_3O^+ + A^-$$

the proton is completely transferred from the acid to the base water. When dealing with rates, however, a transition state of proton transfer is involved. In this transition state the proton is not completely but only partly transferred. The constant α can be regarded as a measure of proton transfer in the transition state. When α is small, e.g., 0.1 or 0.2, in the transition state the proton is transferred only slightly from acid, HA, to substrate, C. When α has a high value, e.g., 0.8 or 0.9, in the transition state the proton is transferred to a large extent. Similarly, a large value of the base catalytic constant β [Eq. (7.2)] indicates appreciable proton transfer in the transition state from substrate to base.

A--H-----C

A------H--C

B--H------C

In an investigation of the mechanisms of such reactions it is important to know whether a reaction is specific or general acid catalyzed. The rate equation is fully elucidated in order to answer this question. Thus, when the rate equation involves both the concentration of the hydronium ions as well as the concentrations of one or more additional acids, the operation of a mechanism of general acid catalysis is confirmed, for example,

$$v = (k_{H_3O^+} c_{H_3O^+} + k_{HOAc} c_{HOAc}) c_{substrate}$$

When the hydronium ions are the only acid species to appear in the rate equation, there are two possibilities:

1. The reaction is subject to specific catalysis.
2. The reaction is general acid catalyzed, but the catalytic coefficients of all acids other than H_3O^+ are too small to be observed. For instance, in the equation

   $$k_1 = k_{H_3O^+} c_{H_3O^+} + k_{RCOOH} c_{RCOOH} + k_{H_2O} c_{H_2O}$$

   when the magnitude of the first term is very large in aqueous solution the last two terms could be too small to be observed. This situation usually occurs when the Brønsted coefficient α is large ($\alpha \rightarrow 1$).

A more general expression of Brønsted's equation is

$$\frac{k_i}{p} = G\left(\frac{qK_{a_i}}{p}\right)^{\alpha} \qquad (7.4)$$

in which p is the number of dissociable protons in the acid molecule and q the number of equivalent positions to which a proton can be bound in the corresponding base: for H_3O^+, p = 3; for H_2O, p = 2; and for CH_3COOH, p = 1; for H_2O, q = 1; for OH^-, q = 1; and for CH_3COO^-, q = 2.

REACTIONS IN STRONG ACIDS

Reactions which take place in strongly acidic medium (from approximately 1 mol/liter and up) can be roughly divided into two groups. The reactions of one group exhibit rates proportional to acid concentration, whereas the reactions of the other group show rates which more or less follow the Hammett acidity function. In the latter group the rate increases much faster than the acid concentration, for example, the acid catalyzed hydrolysis of two esters.

<u>The hydrolysis of α-glyceryl monobenzoate in water at 90°C catalyzed by perchloric acid</u>:

C. T. Chmiel and F. A. Long, J. Amer. Chem. Soc. 78:3326 (1956).

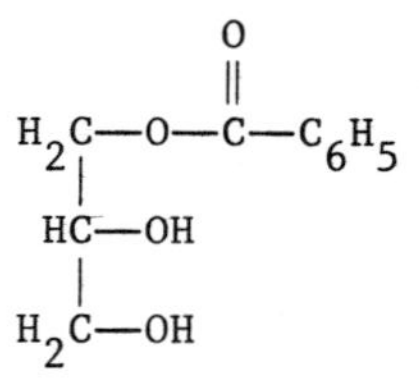

Conc. $HClO_4$ (mol/liter)	$k_1 \times 10^5$ (s^{-1})	$k_2 \times 10^4$ (liters mol^{-1} s^{-1})
0.0446	0.68	1.52
0.223	3.37	1.51
0.926	13.6	1.47
3.01	44.6	1.48
5.77	85	1.47

$$v = k_2 c_{ester} c_{H^+} \qquad k_1 = k_2 c_{H^+}$$

The rate constant k_2 is constant over a large range of acidity.

The hydrolysis of 2,4,6-trimethylbenzoic acid methyl ester has also been determined in aqueous perchloric acid solution at 90°C. The following rate constants were determined:

Conc. $HClO_4$ (mol/liter)	$k_1 \times 10^5$ (s^{-1})	$k_2 \times 10^4$ (liters mol^{-1} s^{-1})
1.00	0.073	0.0073
1.93	0.42	0.022
2.96	1.8	0.061
4.39	12.7	0.29
5.76	114	2.0

In this reaction k_2 is far from being constant; k_1 increases much more rapidly than the acid concentration.

The difference in behavior of these substrates toward the same acid species corresponds to a difference in mechanism. The acid catalyzed hydrolysis of the glyceryl monobenzoate follows the regular pattern of most carboxylic acid esters; see below and Fig. 7.2. The rate determining step is somewhere between II and VI. This means that a water molecule is covalently bound in the slow step.

C. T. Chmiel and F. A. Long, J. Amer. Chem. Soc. 78:3326 (1956).
M. L. Bender, H. Ladenheim, and M. C. Chen, J. Amer. Chem. Soc. 83:123 (1961).

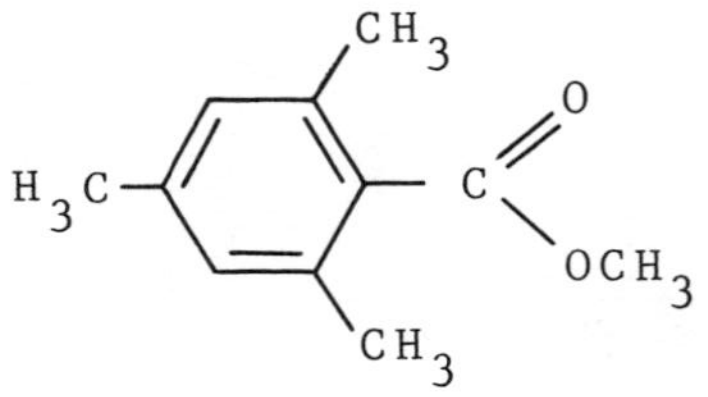

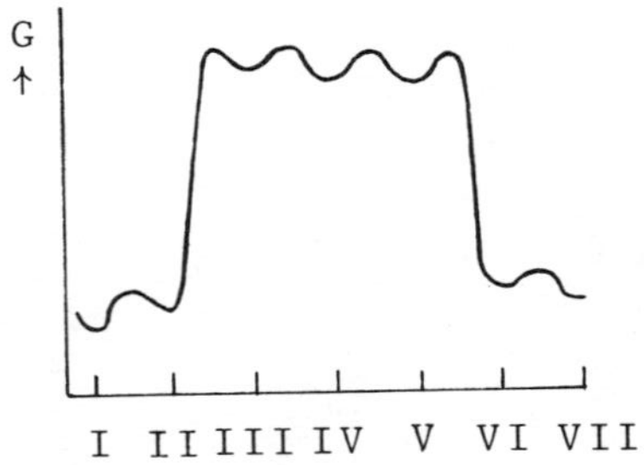

FIG. 7.2

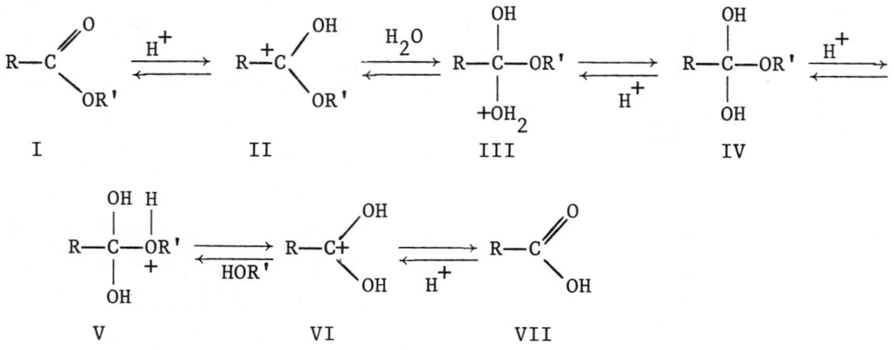

In the methyl ester of 2,4,6-trimethylbenzoic acid the steric effect of both ortho substituents prevents the conversion of the trico-ordinate carbon center in

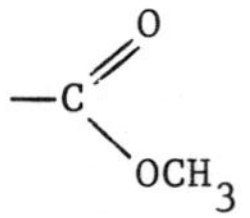

to the tetracoordinate in the orthoacid intermediate

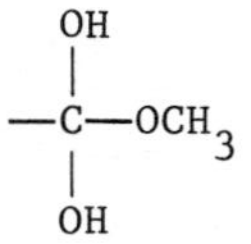

The reaction takes the following path instead:

$$R-C\overset{O}{\underset{\overset{+}{O}R'}{<}} + H^+ \underset{}{\overset{K}{\rightleftharpoons}} R-C\overset{O}{\underset{\overset{+}{O}R'}{<}}\quad\text{rapid (far to the left)}$$

$$\underset{\text{I}}{} \qquad \underset{\text{II}}{\overset{H}{}}$$

$$R-C\overset{O}{\underset{\overset{+}{O}R'}{<}} \overset{k}{\longrightarrow} R-\overset{+}{C}=O + HOR' \quad\text{slow}$$

$$\underset{\text{II}}{\overset{H}{}}$$

$$R-\overset{+}{C}=O + H_2O \longrightarrow R-C\overset{O}{\underset{\overset{+}{O}H_2}{<}} \longrightarrow RCOOH + H^+ \quad\text{rapid}$$

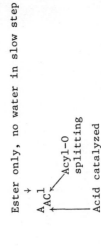

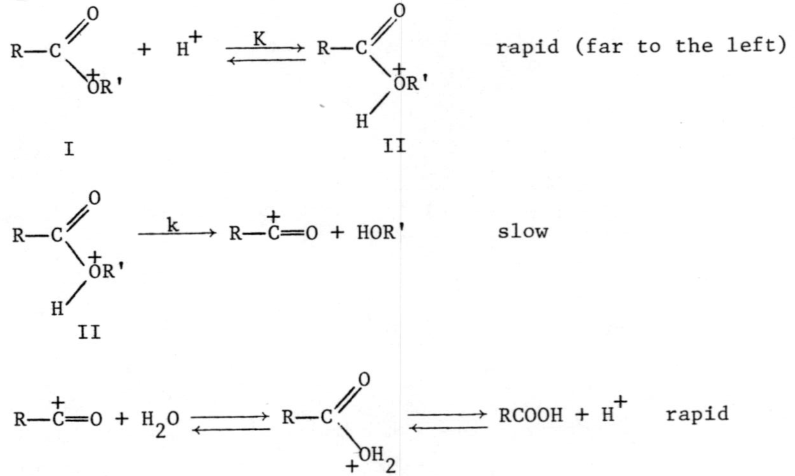

The latter mechanism is named with the abbreviation $A_{AC}1$, and the former, the regular mechanism, $A_{AC}2$. The question is, Why does the rate constant k_2 in the $A_{AC}1$ process increase with acid concentration? One would expect the rate to be

$$v = kc_{II} = kKc_{I}c_{H^+}$$
$$\underbrace{\phantom{kKc_{I}}}_{k_2}$$

and k_2 to be, therefore, independent of acid concentration. The reason for the difficulty is the neglect of activity coefficients, which

Ester only, no water in slow step

$$A_{AC}1$$
$$A_{AC}1 \left\{\begin{array}{l}\text{Acyl-O}\\\text{splitting}\\\text{Acid catalyzed}\end{array}\right.$$

creates error in estimating the effective acid concentrations in concentrated ionic media. When taking activities into account a new parameter arises for consideration, namely, the Hammett acidity function. First we shall discuss how this function has been defined and then how it is measured.

M. A. Paul and F. A. Long, Chem. Rev. 57:1,935 (1957).

HAMMETT ACIDITY FUNCTION

When dissolving a base in an acidic medium, we have the equilibrium

$$B + H^+ \rightleftarrows HB^+$$

with the dissociation constant

$$K_a = \frac{a_B a_{H^+}}{a_{HB^+}} = \frac{c_B}{c_{HB^+}} c_{H^+} \frac{f_B f_{H^+}}{f_{HB^+}}$$

By definition, in dilute aqueous solution all activity coefficients are equal to 1 and thus

$$K_a \frac{c_B}{c_{HB^+}} c_{H^+} \qquad \text{or} \qquad -\log K_a \equiv pK_a = \log \frac{c_{HB^+}}{c_B} + pH \qquad (7.5)$$

When the base is an indicator, e.g., p-nitroaniline, c_B and c_{HB^+} can be determined colorimetrically. The pH can be measured, and from these data pK is known. However, when the same indicator is dissolved in a strong acid solution, e.g., 1 mol/liter of H_2SO_4, the activity coefficients are no longer equal to one ($f \neq 1$), and their actual magnitudes must be taken into account:

$$K_a = \frac{c_B}{c_{HB^+}} \frac{a_{H^+} f_B}{f_{HB^+}} \quad \text{or} \quad \log \frac{c_{HB^+}}{c_B} = pK_a + \log \frac{a_{H^+} f_B}{f_{HB^+}} \qquad (7.6)$$

Since for this indicator pK_a is known and $\log(c_{HB^+}/c_B)$ again can be measured colorimetrically (see Fig. 7.3), the term $\log(a_{H^+} f_B/f_{HB^+})$ can be determined in this concentration range. This new parameter, which takes the place of $\log c_{H^+}$, is denoted by $-H_0$, and H_0 is called the Hammett acidity function; $(a_{H^+} f_B/f_{HB^+})$ is denoted by h_0, and thus

$$H_0 = -\log h_0 \qquad (7.7)$$

$$\log \frac{c_{HB^+}}{c_B} = pK_a - H_0 \qquad (7.8)$$

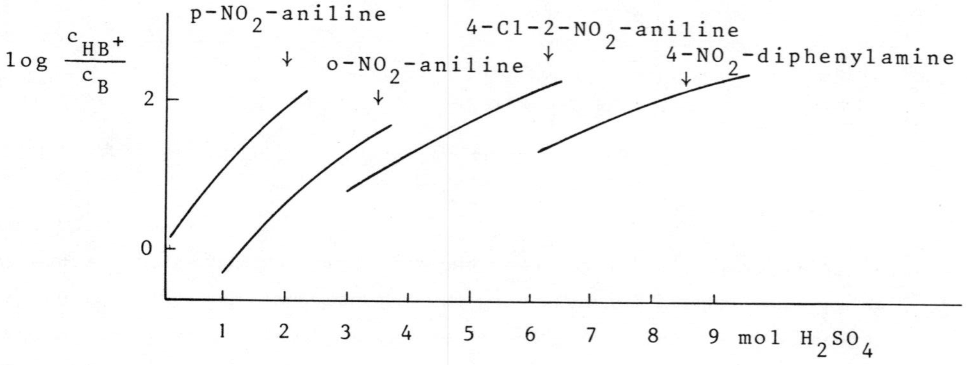

FIG. 7.3

With p-nitroaniline we can increase the acidity up to 2 mol/liter of H_2SO_4. At this acidity it is almost completely in the protonated form HB^+. We now change to a weaker base, e.g., o-nitroaniline. The analogous equation for this indicator is

$$\log \frac{c_{HB^+}}{c_B} = pK_a' - H_0$$

Notice the constant distance between two curves in the overlap region.

In the region from 1 to 2 mol/liter of H_2SO_4, where both indicators can be measured, we know H_0 from the measurements on p-nitroaniline. Thus, pK_a' for o-nitroaniline can be calculated and subsequently be used for the determination of H_0 in the region from 2 to 3 mol/liter of H_2SO_4. We can proceed with still weaker bases.

In this derivation the ratio f_B/f_{HB^+} is required to be independent of the indicator B used. Apparently, this requirement is fulfilled in the series of closely related nitroanilines. In this way the following values for H_0 were obtained for aqueous solutions at 25°C:

mol/liter:	1	2	3	4	5	6	7	8
H_2SO_4:	-0.26	-0.84	-1.38	-1.85	-2.28	-2.76	-3.32	-3.87
$HClO_4$:	-0.22	-0.78	-1.23	-1.72	-2.23	-2.84	-3.61	-4.33

In Eqs. (7.7) and (7.8), H_0 and h_0 have replaced pH and hydronium ion concentration, respectively. The h_0 value increases much more rapidly than acid concentration, as shown in the following example:

$[HClO_4]$ (mol/liter):	1	2	3	4	5	6	7	8
h_0:	1.7	6.0	17	52	170	692	4,070	21,400

$A_{AC}1$ AND $A_{AC}2$ MECHANISMS

We now return to consideration of the rates of reactions observing either the $A_{AC}1$ or $A_{AC}2$ patterns of mechanism, usually referred to as the A1 and A2 alternatives. Activity coefficients must be employed both with equilibria and reaction rates when reactions are taking place in the same acidic media. According to transition-state theory [Eq. (4.38)], for the reaction

$$A + B + \cdots \xrightarrow{k_r} \text{products}$$

we have

$$k_r = (k_r)_0 \frac{f_A f_B \cdots}{f^{\neq}}$$

where $(k_r)_0$ is the rate constant in ideal dilute aqueous solution, or, in any case, in a medium where the activity coefficients are assumed to be 1.

<u>For the A1 mechanism</u>:

(1) $S + H^+ \overset{K}{\rightleftharpoons} HS^+$ rapid equilibrium far to the left

(2) $HS^+ \xrightarrow{k} X^+ + Y$ slow step

(3) $X^+ + H_2O \longrightarrow \text{products} + H^+$ rapid subsequent reaction

$$\left.\begin{aligned} v &= kc_{HS^+} = k_0 c_{HS^+} \frac{f_{HS^+}}{f^{\neq}} = k_0 \frac{a_{HS^+}}{f^{\neq}} \\ K &= \frac{a_{HS^+}}{a_S a_{H^+}} \end{aligned}\right\} \begin{aligned} v &= k_0 K a_S a_{H^+} \frac{1}{f^{\neq}} \\ &= k_0 K c_S \frac{a_{H^+} f_S}{f^{\neq}} \end{aligned} \qquad (7.9)$$

$B + H^+ \rightleftarrows HB^+$

$S + H^+ \rightleftarrows HS^+$

When v is written as $v = k_1 c_S$, the first-order rate constant $k_1 = k_0 K(a_{H^+} f_S/f^{\neq})$. The term $(a_{H^+} f_S/f^{\neq})$ is similar to $h_0 = a_{H^+} f_B/f_{B^+}$. The transition-state complex $HS^{\neq}$ resembles HS^+. Therefore, it is probable that $f_S/f^{\neq}$ is equal to or proportional to f_B/f_{HB^+}, and thus, $a_{H^+} f_S/f^{\neq} = \text{constant} \times h_0$:

$$k_1 = \text{constant}' - H_0$$

or

$$\log k_1 = \text{constant}'' - h_0 \qquad (7.10)$$

According to this reasoning, for an A1 reaction a plot of log k_1 versus $-H_0$ must give a straight line with slope +1. And indeed, when the log k_1 in the hydrolysis of methyl 2,4,6-trimethylbenzoate is plotted versus $-H_0$, full linear behavior is observed. See Fig. 7.4. The only discrepancy is that the slope of the line is not exactly equal to 1. This discrepancy originates with the approximations made in deriving the equation applied above.

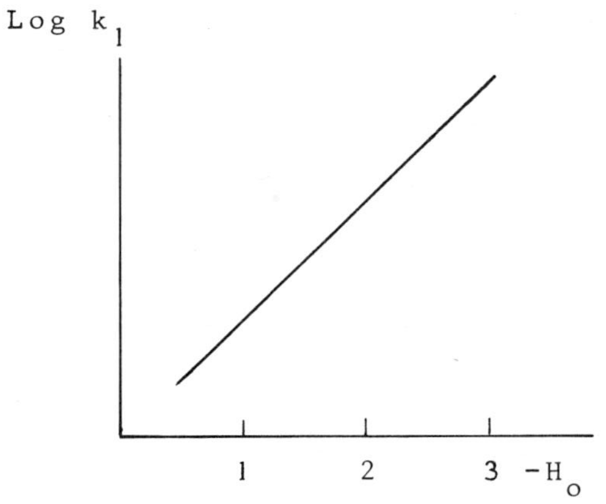

FIG. 7.4

In the A1 mechanism, log k_1 is linear in H_0, and k_1 is approximately proportional to h_0. According to the mechanism on page 179, for an A2 reaction k_1 would be proportional to

$$c_{H^+} \frac{f_{H^+} f_S a_{H_2O}}{f^{\neq}}$$

Empirically, k_1 appears to be proportional to $c_{H_3O^+}$. Apparently, the term $(f_{H^+} f_S a_{H_2O})/f^{\neq}$ is independent of the acidity.

ZUCKER-HAMMETT AND BUNNETT CRITERIA

The linear relation between log k_1 and $-H_0$ is sometimes regarded as an indication of the absence of a covalently bound H_2O molecule in the transition state of the slow step. On the other hand, when log k_1 is linear in $c_{H_3O^+}$, a water molecule would be covalently present. This criterion is called the Zucker-Hammett criterion.

For details about this and other acidity functions, see C. H. Rochester, Acidity Functions (Academic, London, 1970).

After its introduction in 1939 many exceptions to this classification have been observed. In 1960 Bunnett introduced another criterion, based on a plot of log $k_1 + H_0$ versus log a_{H_2O}, where a_{H_2O} is the activity of the water in the aqueous acid solution. In most cases this plot is approximately a straight line, slope w. Mainly on empirical grounds, Bunnett made the following classification:

J. F. Bunnett, J. Amer. Chem. Soc. 82:499 (1960); 83:4956-4978 (1961).

w	
-2.5 to 0.0	H_2O not covalently present in the transition state of the slow step
+1.2 to +3.3	H_2O present as a nucleophile
> 3.3	H_2O acts as a proton transfer agent

8 Chain Reactions

Distinctive kinetic features of commonly encountered free radical chain processes

An example of a chain reaction, namely, the formation of HBr from H_2 and Br_2 in the dark at 300°C in the gas phase, has been mentioned on p. 33. In this example, the process is *initiated* by the dissociation of a bromine molecule; the *chain carriers* are the $Br^\bullet$ and $H^\bullet$ atoms; in the *propagation steps* these carriers react with H_2 and Br_2, respectively; and the chain is *terminated* most commonly when two bromine atoms recombine.

Many chain processes are encountered in the field of polymer formation. Vinyl polymerization is one of the most widely applied reactions in chemical industry. Most of our understanding of the course of such reaction processes and their chain characteristics stems from kinetic studies. An instructive illustration is provided by the peroxide or photoinduced polymerization of monomers such as styrene or vinyl acetate. The lines of evidence pointing to very long kinetic chains in these reactions are threefold:

1. In the absence of any species which can inhibit or regulate polymer chain growth these polymerizations lead to very high molecular weight products.
2. Only a minute proportion of the peroxide initiator suffers thermal decomposition during the course of reaction, indicating that only a small number of chain-initiating reactions are required to consume the larger proportion of monomer. Where a photoinitiator is employed a similar conclusion may be deduced.
3. Increasing the concentration of peroxide initiators, which experience unimolecular, thermal decomposition, or increasing the intensity of photoinitiation produces a decline in the chain lengths. This is taken as an indication of the fact that the higher the concentration of chain-carrying species in solution, the higher the rate of mutual annihilation through bimolecular recombination or disproportionation.

SUBSEQUENT STEPS

These observations are in accord with the following mechanistic events when initiation is by peroxide (P) decomposition and termination is by radical recombination. The concentration of the monomer, $CH_2{=}CHX$, is denoted by [M].

Initiation: $$P \xrightarrow{k_i} 2R^{\cdot}$$

Propagation: $$R_n^{\cdot} + M \xrightarrow{k_p} R_{n+1}^{\cdot}$$

Termination: $2R^{\cdot} \xrightarrow{k_t} \text{polymer}$

KINETIC EQUATIONS

We make the reasonable assumption that all radicals with the same terminal structure $\sim\!\sim CH_2—\dot{C}HX$ have the same reactivity in propagation and in termination. Application of the steady-state assumption to the radical concentration gives

rate of initiation = rate of termination

or

$$2k_i[P] = 2k_t[R^{\cdot}]^2 \tag{8.1}$$

$$[R^{\cdot}] = \left(\frac{k_i[P]}{k_t}\right)^{1/2} \tag{8.2}$$

For the propagation reaction

$$\frac{-d[M]}{dt} = k_p[R^{\cdot}][M]$$

and after integration

$$\ln\frac{[M_0]}{[M]} = \left(\frac{k_p^2 k_i[P]}{k_t}\right)^{1/2} t \tag{8.3}$$

INHIBITION

Equation (8.3) describes a linear relation between a logarithmic function of the monomer concentration (initial value $[M_0]$) and the time of reaction. If we add a substance to the reaction, often called an inhibitor (Inh), which can intercept the radical $R^\cdot$ and prevent much chain growth, we are in effect introducing another chain-terminating step:

$$R^\cdot + \text{Inh} \xrightarrow{k_{\text{Inh}}} \text{inert product}$$

There are several possible kinetic developments concerned with the various possibilities stemming from the relative efficiency of Inh as an inhibitor. Thus, if Inh is a very efficient inhibitor, it could sweep up radicals $R^\cdot$ so effectively that little or no $R^\cdot + R^\cdot$ termination could occur while Inh is present. This behavior would constitute one extreme type of inhibition.

A second type would entail an inhibiting reagent which attacks $R^\cdot$ with a degree of effectiveness comparable to that with which chain-carrying radicals annihilate each other. The presence of this type of initiator would be manifested by a reduced rate of polymerization compared to what is characteristic of the pure monomer at the stipulated peroxide initiator levels.

Such a reduction of rate is present in the first part of curve b in Fig. 8.1.

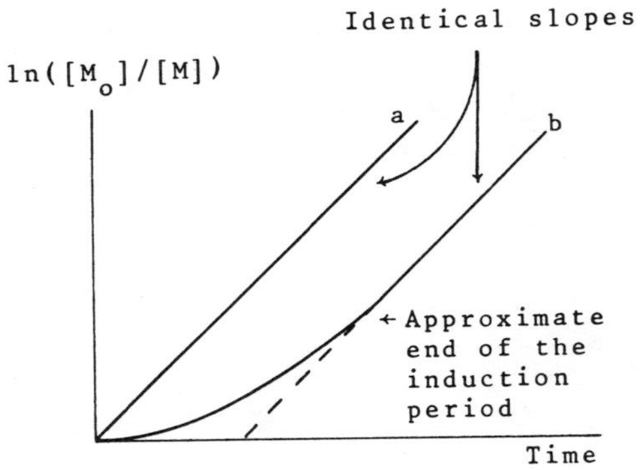

FIG. 8.1. The function $\ln([M_0]/[M])$ for vinyl acetate plotted vs. time. Curve a: without inhibitor; curve b: with duroquinone (tetramethyl-1,4-benzoquinone), a very efficient inhibitor.

INDUCTION PERIOD

The assumption that no bimolecular $R^{\cdot} + R^{\cdot}$ termination occurs holds nearly to the point where [Inh] is exhausted. The interval of time up to this period is that which is required to attain the steady-state rate characteristic of the uninhibited reaction and is called the induction period. This has been well verified in cases where both the rate of disappearance of Inh and the rate of formation of initiating radicals in solution have been followed simultaneously by independent analytical methods. The assumption is made that the peroxide concentration, [P], is constant, $[P_0]$, and that the rate of radical formation is equal to that of the unimolecular rate of peroxide decomposition.

Let y be the fraction of radicals $R^{\cdot}$ capable of reacting with Inh, i.e., the chain-stopping efficiency of Inh. The steady-state assumption leads to (rate of formation) = (rate of termination):

$$2k_i[P_0] = yk_{Inh}[R^{\cdot}][Inh] \tag{8.4}$$

$$[R^{\cdot}] = \frac{2k_i[P_0]}{yk_{Inh}[Inh]} \tag{8.5}$$

If $2/y = a$ and since (propagation step, p. 190) $-d \ln[M]/dt = k_p[R^{\cdot}]$, we get

$$\frac{-d \ln[M]}{dt} = \frac{ak_pk_i[P_0]}{k_{Inh}[Inh]} \tag{8.6}$$

The rate of disappearance of Inh is given by

$$\frac{d[\mathrm{Inh}]}{dt} = -yk_{\mathrm{Inh}}[R^{\bullet}][\mathrm{Inh}]$$

$$= -2k_i[P_0] \qquad [\text{Eq. } (8.4)] \tag{8.7}$$

Integration gives

$$[\mathrm{Inh}] = [\mathrm{Inh}_0] - ak_i[P_0]t \tag{8.8}$$

where $[\mathrm{Inh}_0]$ is the initial inhibitor concentration. On substituting this value of Inh into the expression for the monomer consumption rate (8.6) we get

$$\frac{-d\ \ln[M]}{dt} = \frac{k_p}{k_{\mathrm{Inh}}} \frac{ak_i[P_0]}{[\mathrm{Inh}_0] - ak_i[P_0]t} \tag{8.9}$$

or its reciprocal

$$\left(\frac{-d\ \ln[M]}{dt}\right)^{-1} = \frac{k_{\mathrm{Inh}}}{k_p} \frac{[\mathrm{Inh}_0]}{ak_i[P_0]} - \frac{k_{\mathrm{Inh}}}{k_p} t \tag{8.10}$$

Integration of (8.9) results in

$$\ln\left(\frac{[M_0]}{[M]}\right) = \frac{-k_p}{k_{\mathrm{Inh}}} \ln \frac{[\mathrm{Inh}_0] - ak_i[P_0]t}{[\mathrm{Inh}_0]} \tag{8.11}$$

The values of $d \ln[M]/dt = (d[M]/dt)/[M]$ can be determined from the tangent of each point on the polymerization curve. According to Eq. (8.10), a plot of $(-d \ln[M]/dt)^{-1}$ versus t affords a straight line. The slope of this line gives us k_{Inh}/k_p. From the intercept, which is

$(k_{Inh}/k_p)([Inh_0]/ak_i[P_0])$, and knowledge of $[Inh_0]$, $[P_0]$, and a, we can check any previously measured values of k_i, the unimolecular decomposition rate of peroxide. Or, if we know k_i, we can use this knowledge to calculate $a = 2/y$, where y is the chain-stopping efficiency of an inhibitor molecule Inh.

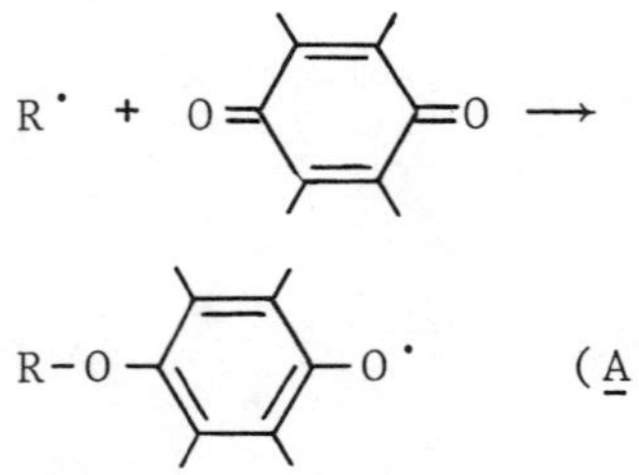

(A)

Dead radical, not an inhibitor

The preceding equations have been applied to the inhibitor duroquinone for which y appears to be unity. If the product of the inhibition reaction (A) were also an inhibitor, then y would be found to be 2.

This quinone and other efficient inhibitors may also be applied for determination of the initiation rate in photoinitiated polymerization. For this purpose Eq. (8.1) and subsequent equations are altered by substitution of the initiation term f(I) for $2k_i[P_0]$. Here and in the following equations, I is the photoinitiator.

Returning to the situation where an inhibitor is absent, the photoinduced vinyl polymerization is characterized by the steps

Initiation: $I \xrightarrow{h\nu} 2R^{\cdot}$

Propagation: $R_n^{\cdot} + M \xrightarrow{k_p} R_{n+1}^{\cdot}$

The assumption can be justified that during the first few percent of conversion k_p is equal for all radicals in solution.

There are two typical modes of chain ending or termination:

1. Disproportionation:

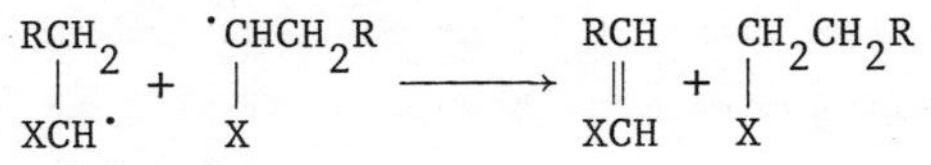

$$\begin{array}{ccc}
RCH_2 & \overset{\cdot}{C}HCH_2R \\
\overset{|}{XCH\cdot} & \overset{|}{X}
\end{array}
\longrightarrow
\begin{array}{ccc}
RCH & CH_2CH_2R \\
\overset{||}{XCH} & \overset{|}{X}
\end{array}$$

where the polymer chain lengths are equal to the kinetic chain lengths

2. Dimerization:

$$R\cdot + R\cdot \longrightarrow R—R$$

where the polymer chain lengths are twice as great as the kinetic

It must also be clear that

1. k_p is a measure of the reactivity of $R\cdot$ toward monomer.
2. The magnitude of k_p is related to the stability of $R\cdot$ and is controlled by the substituent X in the monomer $CH_2{=}CHX$, where X = Ph, Cl, OAc, etc.
3. k_t is larger when $R\cdot$ is less stable.

Thus, if we can measure k_p and k_t, we can attain a direct estimate of radical stability as a function of structure. How do we measure k_p and k_t? The following kinetic equations can be deduced from the mechanistic steps depicted above:

Monomer consumption: $-d[M]/dt = k_p[R\cdot][M]$ (8.12)

Variation in $R\cdot$: $d[R\cdot]/dt = f[I] - 2k_t[R\cdot]^2$ (8.13)

STEADY-STATE RATE CONSTANT

Assuming that variation in $[R^\cdot]$ is very small during the course of kinetic observation, a steady-state treatment can be applied. Therefore, $d[R^\cdot]/dt \simeq 0$, and $f[I] = 2k_t[R^\cdot]^2$.

Thus, $[R^\cdot] = (f[I]/2k_t)^{1/2}$, and, on substituting into Eq. (8.12), $-d(\ln[M])/dt = k_p(f[I]/2k_t)^{1/2}$, which integrates to

$$\ln \frac{[M_0]}{[M]} = \left(\frac{k_p^2}{2k_t} f[I]\right)^{1/2} t \tag{8.14}$$

Thus, the steady-state rate constant k_s is given by

$$k_s = \left[\frac{k_p^2}{2k_t} f(I)\right]^{1/2} \tag{8.15}$$

and is comparable to the rate constant of a first-order equation.

The experimental results realized by following the concentration of monomer as a function of time completely justify the steady-state assumption for $R^\cdot$. Thus, as shown in Fig. 8.2, the straight-line relationship passing through the origin is obeyed from the very first moments of reaction up to about 7% conversion, when further observation is halted due to a great increase in medium viscosity. If any significant amount of time were required for the concentration of $R^\cdot$ to build to the steady-state value, the curved dashed line would have been expected.

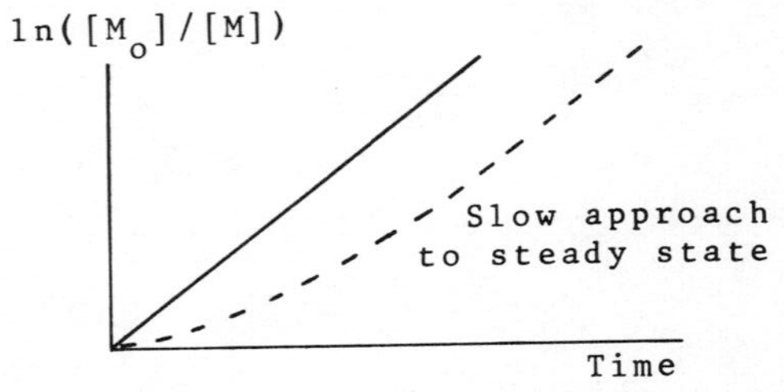

FIG. 8.2. Plot of $\ln([M_0]/[M])$ vs. time. The slope of the straight line is equal to $k_s = \{(k_p^2/2k_t)f[I]\}^{1/2}$.

Two important conclusions are to be derived from this exercise:

1. The concentration of the chain-carrying radical species is ordinarily very small, and it builds to its steady-state value very rapidly.
2. The steady-state assumption imposes a limitation in that only relative values of the absolute rate constants of propagation (k_p) and termination (k_t) can be gained, since knowing the initiation rate f[I] only permits calculation of $k_p^2/2k_t$ from the steady-state slope.

Absolute values of the rate constants k_t and k_p can be obtained by application of the rotating sector method.

ROTATING SECTOR METHOD

This method can be readily illustrated by the determination of k_p and k_t in the photoinitiated polymerization of vinyl acetate where the steady-state assumption in $R^\cdot$ ordinarily applies from the very beginning of reaction. This method has also been applied to autoxidation reactions by Ingold and co-workers. The rationale of the rotating sector measurement stems from the non-steady-state equations summarized as follows:

J. A. Howard and K. U. Ingold, Can. J. Chem. 43:2729 (1965); 44:1119 (1966).

Constant initiation rate (producing $2R^\cdot$): $$f[I] \tag{8.16}$$

Polymerization rate: $$\frac{-d\ \ln[M]}{dt} = k_p[R^\cdot] \tag{8.17}$$

Termination rate: $$\frac{d[R^\cdot]}{dt} = -2k_t[R^\cdot]^2 \tag{8.18}$$

Variation in $R^\cdot$: $$\frac{d[R^\cdot]}{dt} = f[I] - 2k_t[R^\cdot]^2 \tag{8.19}$$

[Equation (8.19) ≡ Eq. (8.13).]

When we turn on the beam of light the entire face of the reaction cell is uniformly illuminated. This beam is weakly absorbed by the photoinitiator and thereby affords relatively uniform initiation of polymerization throughout the cell. The curve in Fig. 8.3 is a plot of the growth in radical concentration in the first few moments of such illumination.

If the illumination is not interrupted, the concentration of $R^{\cdot}$ soon attains the steady-state value $[R^{\cdot}_s]$. If we turn off the light at some point t' prior to attainment of $[R^{\cdot}_s]$, the radical concentration decays along a curve declining from point t' according to the integral of the equation

$$\frac{d[R^{\cdot}]}{dt} = -2k_t[R^{\cdot}]^2 \quad \text{or} \quad \frac{1}{[R^{\cdot}]} = 2k_t t + \text{constant} \tag{8.20}$$

The reciprocal is

$$[R^{\cdot}] = \frac{1}{2k_t t + \text{constant}} \tag{8.21}$$

A pie-shaped disc is used to intercept the beam and thus to continue this cycle of light and dark in alternation. See Fig. 8.4. The illuminated fraction of the total cycle of light and dark is equal to $1/(q + 1)$, where the dark period = $q \times$ (the light period). The radical concentration as a function of time is shown in Fig. 8.5.

The average radical concentration is equal to the area under the curve, the shaded area, divided by the time base $t_3 - t_1$. For various

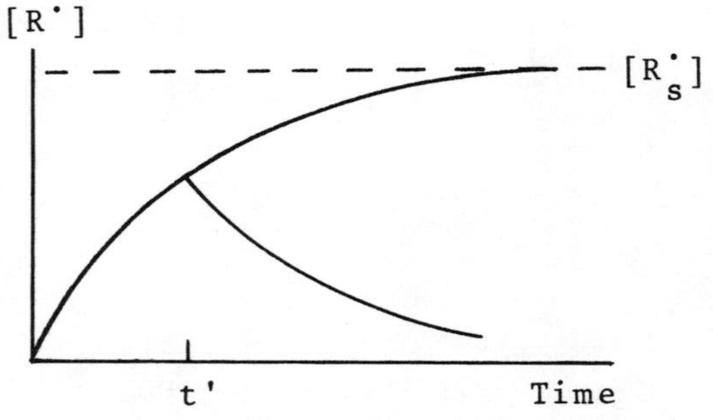

FIG. 8.3. Radical concentration as a function of time. $[R^{\cdot}_s]$ is the steady-state value of $[R^{\cdot}]$.

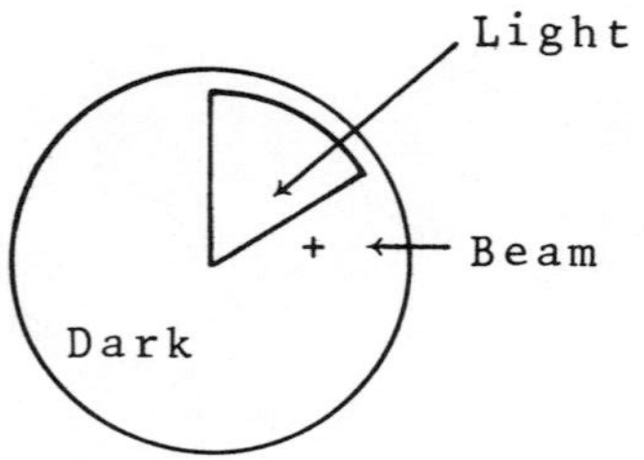

FIG. 8.4. Rotating sector. The beam is perpendicular to the plane.

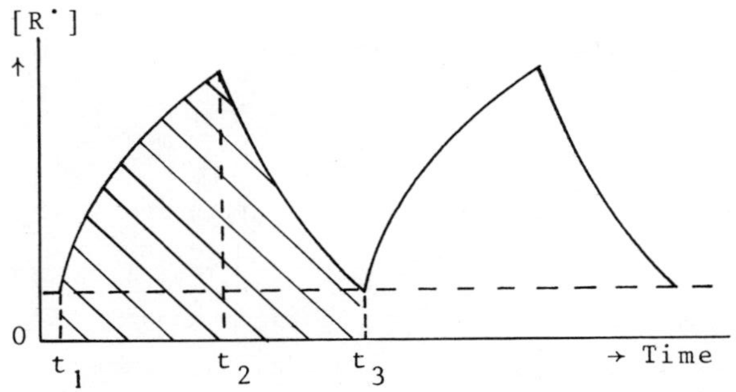

FIG. 8.5. Radical concentration as a function of time. From t_1 to t_2, illumination; from t_2 to t_3, darkness.

frequencies of flashing, different areas can be anticipated. Mathematical equations to calculate the size of the shaded area can be derived as follows.

Under illumination conditions (period $t_1 \rightarrow t_2$), Eq. (8.13) ≡ (8.19) applies. Integration of it gives

$$\frac{1}{\sqrt{2k_t}} \frac{1}{\sqrt{f[I]}} \tanh^{-1} \frac{\sqrt{2k_t}\,[R^\cdot]}{\sqrt{f[I]}} = t + \text{constant}$$

$$\frac{\sqrt{2k_t}\,[R^\cdot]}{\sqrt{f[I]}} = \tanh\{\sqrt{2k_t}\,\sqrt{f[I]}\,(t + \text{constant})\}$$

See any table of integrals.

$$\int \frac{dx}{a^2 - x^2} = \frac{1}{a} \tanh^{-1} \frac{x}{a}$$

$x = \sqrt{2k_t}\,[R^\cdot]$ and $a = \sqrt{f[I]}$

$$[R^\cdot] = \left(\frac{f[I]}{2k_t}\right)^{1/2} \tanh\left\{(2k_t f[I])^{1/2}(t + \text{constant})\right\} \quad (8.22)$$

During the period $t_2 \rightarrow t_3$ the decay of radical concentration is expressed by Eq. (8.20).

We simplify Eqs. (8.22) and (8.20) by defining two new variables:

$$\tau = \beta t \quad (8.23)$$

$$\gamma = \frac{[R^\cdot]}{[R_s^\cdot]} \quad (8.24)$$

where β is an abbreviation for $(2k_t f[I])^{1/2}$ and $[R_s^\cdot] = (f[I]/2k_t)^{1/2}$; see p. 196.

$[R_s^\cdot]$ is the radical concentration at steady-state condition. It follows from Eqs. (8.13) ≡ (8.19) that for this condition

$$0 = f[I] - 2k_t[R_s^\cdot]^2$$

from which the expression for $[R_s^\cdot]$ is derived.

By substituting the new variables τ and γ, Eqs. (8.22) and (8.20) simplify to (8.25) and (8.26), respectively:

$$\gamma = \tanh(\tau + C) \quad (8.25)$$

$$\gamma = \frac{1}{\tau + C'} \quad (8.26)$$

Two plots are made. One of them corresponds to the equation

$$\gamma = \tanh \tau$$

and the other to

$$\gamma = \frac{1}{\tau}$$

We do not know the constants C and C'. In other words, we can slide the curves over each other along the τ-axis. The time periods $t_1 \rightarrow t_2$ and $t_2 \rightarrow t_3$ in Fig. 8.5 correspond to the intervals τ_1 and τ_2 for the tanh τ and the $1/\tau$ curves, respectively. We do not yet know the absolute values of τ_1 and τ_2, but we know their ratio: $\tau_2 = q\tau_1$. For each value of τ_1, there is one value for τ_2 ($= q\tau_1$) and only one position of the curves where they obey the condition that their highest and lowest values have to be equal. For each such positioning of the superposed curves a shaded area is subtended (cf. Fig. 8.5) which can be readily determined. Each such total area divided by the appropriate value of $\tau_1 + \tau_2$ gives the average value of γ, i.e., $\bar{\gamma}$. In Fig. 8.6 $\bar{\gamma}$ is plotted as a function of $\log(\tau_2 - \tau_1)$.

The plotted quantity $\bar{\gamma} = [R^\cdot]/[R_s^\cdot]$ cf. (8.24) . Assuming that the rate of polymerization is proportional to the radical concentration, $\bar{\gamma}$ is equal to the ratio of the rates under flashing and steady-state conditions; $\bar{\gamma}$ = rate (flashing)/rate (steady state). This "theoretical" curve is now compared with an experimentally obtainable curve of approximately the same shape, Fig. 8.7. The experimental curve is obtained by plotting the polymerization rate at various flashing frequencies against $\log(t_2 - t_1)$. Comparing the midpoints of the two curves, we get corresponding values of τ and t and, thus, their ratio [cf. (8.23)]:

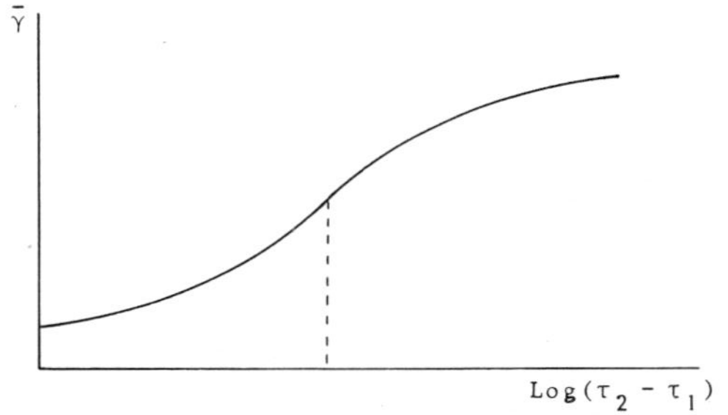

FIG. 8.6 Average value of γ as a function of $\log(\tau_2 - \tau_1)$.

$$\frac{t_3 - t_2}{t_2 - t_1} = q$$

$$\beta = \frac{\tau}{t}$$

where $\beta = (2k_t f[I])^{1/2}$.

Knowing f[I] from an independent measurement of the initiation rate under illumination, we can arrive at a value of k_t. From $[R_s^{\cdot}] = (f[I]/2k_t)^{1/2}$ we are able to calculate the radical concentration under steady-state conditions, $[R_s^{\cdot}]$. Using the rate constant determined with uninterrupted illumination, k_s, and Eq. (8.15), we arrive at the value of k_p.

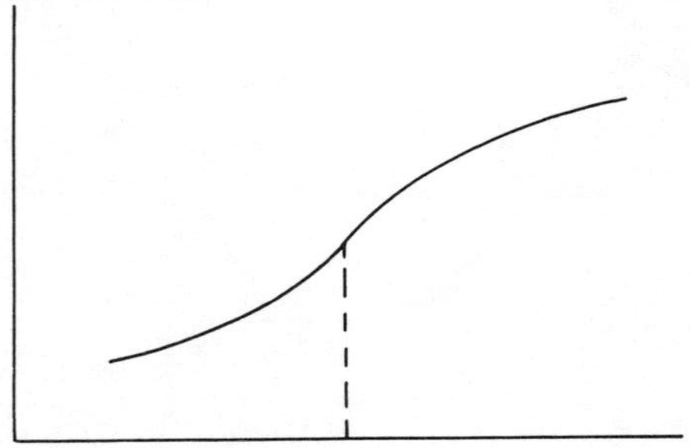

FIG. 8.7. Polymerization rate as a function of $\log(t_2 - t_1)$ in seconds per flashing cycle.

Index